Fixing Intelligence

Fixing Intelligence

FOR A MORE SECURE AMERICA

William E. Odom

Yale University Press   New Haven and London

The book is based principally on a study supported and published by the National Institute for Public Policy in 1997. Both the analysis and several of the recommendations have been updated and significantly modified.

The original study was conceived, chaired, and drafted by Lt. Gen. William E. Odom, U.S. Army, retired.

Members of the study group were: Dr. William Graham, former science adviser to President Reagan; Mr. Robert E. Rich, former deputy director, NSA; Ms. Elisabeth R. Rinkskopf, former legal counsel, NSA and CIA; Lt. Gen. Harry E. Soyster, U.S. Army, retired, former director, DIA; Dr. Gregory Treverton, former chairman, National Intelligence Council; Lt. Gen. James R. Clapper, U.S. Air Force, retired, former director, DIA.

Although they made valuable contributions to and approved the original study, they have not reviewed or approved this book. Thus they bear no responsibility for the changes in the text.

Printed in the United States of America.

Library of Congress Cataloging-in-Publication Data

Odom, William E.

Fixing intelligence : for a more secure America / William E. Odom.

p.   cm.

Includes bibliographical references and index.

ISBN 0-300-09976-2 (cloth : alk. paper)

1. Military intelligence—United States.   2. Intelligence service—United States.

3. United States—Politics and government—2001–.   I. Title.

UB251.U5 O33 2003

355.3'432'0973—dc21   2002015565

A catalogue record for this book is available from the British Library.

The paper in this book meets the guidelines for permanence and durability of the Committee on Production Guidelines for Book Longevity of the Council on Library Resources.

10 9 8 7 6 5 4 3 2 1

Contents

Acknowledgments

Jonathan Brent, the editorial director of the Yale University Press, caused this book to be written. My interest in the topic, fairly intense in the early 1990s, had flagged. Jonathan revived it in the aftermath of 11 September 2001, insisting that the issues of intelligence reform could be made accessible to a wide reading public. I had always been skeptical, but he prevailed, and I decided to try.

If I have succeeded, Dan Heaton, the manuscript editor, deserves much of the credit. He waged a relentless war against acronyms and bureaucratic jargon, winning battle after battle.

The original study on which the book is based owes a lot to Bernard C. Victory, the manuscript editor at the National Institute of Public Policy and my executive assistant who supported the study group.

My biggest debt is to the hundreds of intelligence professionals, officers and enlisted, civilians from the lowest to the highest ranks, who have taught me how U.S. intelligence operations work. Their competence and courage cannot be exaggerated.

Introduction

This book deals with a timely though unpopular topic, intelligence reform. Some members of the Senate gave public attention to reform in the late 1980s and early 1990s, but then their colleagues legislated a presidential commission to investigate the matter. Like most other presidential commissions, its sponsors wanted it to justify the status quo and silence the proponents of reform. They succeeded. Commissions on the National Reconnaissance Office and the National Imagery Agency in 2000–2001 did the same thing in spite of abundant evidence that both needed fundamental changes.

In the aftermath of the events of 11 September 2001, demands for reform began to be taken more seriously. The executive branch remained opposed to change, but public voices and members of Congress were objecting to "business as usual" in light of the greatest intelligence failure since Pearl Harbor. The president's sudden reversal on homeland defense, however—initially opposing a new border control department but then sending draft legislation to Congress for a vastly more complex reorganization, a homeland security department—has complicated the intelligence reform issue, thrusting it to the fore, then deferring it until after creation of this new department.

All of this congressional and presidential maneuvering must not obscure the undeniable: whether to reform U.S. intelligence is no longer the question. Responsible leaders and legislators must now cease their bickering and ask, "What reforms make sense? Will they be effective?"

This book, based in large part on a study published by the National Institute of Public Policy in 1997, should help answer such difficult questions and make it possible for nonexpert observers to distinguish between sense and nonsense in the debate. Moreover, the recommendations that follow offer yardsticks for assessing any changes that Congress or the executive branch may make.

That this has been made possible for nonexperts does not mean that it has been made easy for them. Both the diagnosis of the Intelligence Community's ills and the prescriptions for their cures involve technical language. With the able assistance and prodding of my editors at Yale University Press, however, I have fought a war against acronyms and buzzwords to strip away much of the arcane terminology. Thanks largely to their persistence, I believe the text is reasonably accessible.

The novices and other "lay" readers should know that they are not alone in their ignorance of the labyrinth of U.S. intelligence organizations and activities. Most officials in the intelligence "clergy" are equally clueless but reluctant to admit it. Compartmentalization of organizations and programs—designed to limit knowledge of them and prevent their discovery by hostile intelligence services—makes it possible for a career intelligence official to remain ill-informed, often totally ignorant, of the operations of other offices within his own agency, not to mention the workings of other intelligence agencies throughout the U.S. government. They are like workers in a network of mineshafts, tunnels, and underground caverns.

Rare is the senior intelligence officer who has the overall picture. Moreover, many of them do not even understand their own mineshaft very well because they have grown up in some small tunnel within it. This is understandable when we realize how diverse and alien various intelligence skills and activities can be. Case officers who recruit spies abroad need the skills of a "con artist." Code breakers need to be world-class mathematicians. Between these two extremes are many other required skills that do not mix well, that cultivate different and incompatible cultures.

Talk about reform among senior intelligence officers, therefore, tends to be either incoherent babble or designed to mislead others in order to protect one's parochial organizational interests. With some effort, however, both kinds of facades can be stripped away because the structural arrangements they conceal are not secret. They can be learned from numerous unclassified documents. Nor are the basic processes of the intelligence collection, analysis, and distribution a secret. A clear understanding is publicly available. Moreover, the original study on which this book is based was cleared for publication by the CIA. My purpose here is to make intelligence clear to anyone who truly wants to understand it, including intelligence professionals.

Today's Reform Efforts: What Is at Stake

The events of 11 September 2001 cast a dark shadow over the Intelligence Community. Why was there no intelligence available to warn of the Qaeda attacks on the World Trade Center and the Pentagon? No intelligence failure since December 1941 has been as great. Going back to the events leading up to the attack on Pearl Harbor is instructive today. Neither naval nor army intelligence had warned of an imminent Japanese threat in fall 1941. One of the major reasons given for the creation of the Central Intelligence Agency (CIA) in 1947 was to prevent "another Pearl Harbor." But in June 1950 the CIA failed to warn of North Korea's attack on South Korea. Once again, the United States found itself at war and ill-prepared to fight.

Are there any lessons here for how the United States should respond to its most recent intelligence failure? Indeed there are, but it is easy to learn a false lesson—namely, that reorganizing intelligence makes no sense because it did not prevent an earlier intelligence failure. In fact, reorganization made sense in spite of the failure. The proper lesson is that there are no easy fixes, that the problems are complex and structural, and that they cannot be reduced to television sound bites. This point becomes more compelling if we take a moment to understand why it would have been irresponsible not to have made the 1947 reforms.

Neither the army nor the navy had an effective intelligence system before World War II, although the War Department had created a rather successful signals intelligence unit during World War I. Known by insiders as American Black Chamber, after the war it was jointly funded by the War and State Departments, performing some remarkable code-breaking feats until Henry Stimson became secretary of state in 1929. Briefed on its activities, he ordered it closed because "gentlemen do not open other gentlemen's mail." The Black Chamber had deciphered Japanese diplomatic codes during World War I, allowing it to intercept Tokyo's guidance to its diplomats at the Washington Naval Conference after the war, an advantage U.S. negotiators exploited. The Black Chamber's closure forced its former head into the jobless ranks of the Great Depression. Remarkably talented, this card shark, code breaker, and journalist promptly wrote a book recounting the Black Chamber's many feats.[1] It was a commercial success, especially in Japan, prompting that government to change its codes. Not until the beginning of World War II did the navy's signals intelligence (which Stimson was not in a position to disband) begin to break some of the new ones—but not the right ones and not in time to anticipate the attack on Pearl Harbor. The country had been cruelly served by Stimson's decision.

Military and naval attachés were the only other intelligence collectors of note in either department. Into this yawning gap in military intelligence stepped William O. Donovan, who, having used his access to President Roosevelt to press for some kind of civilian intelligence capability outside the military departments and working directly for the president, was allowed to create the Office of the Coordinator of Information (COI) in June 1941.[2] A winner of the Congressional Medal of Honor and the Distinguished Service Cross in World War I, Donovan was a hero as well as the head of a very successful New York law firm. He used his considerable reputation and political connections to overcome the opposition of the military departments to this new entity, but as its first director, he was unable to assert control of the army and navy intelligence capabilities. A new round of bureaucratic struggles followed the outbreak of war six months later, leading to transformation of the COI into the Office of Strategic Services (OSS) in June 1942. This time, Donovan compromised with the military departments, placing the OSS nominally under the Joint Chiefs. Donovan took command of it as a colonel and was subsequently promoted to brigadier and major general. Although its personnel were in uniform and held military ranks, the OSS was not subject to military direction of its internal organization and personnel policies. The Joint Chiefs finally gave it official responsibility for sabotage, espionage, counterespionage, and covert action in December 1942, but Donovan was already doing some of these things by that time. Although the COI carried over as the Research and Analysis division and remained the heart of the new organization's Washington center, the OSS soon was acting as much like the forerunner of today's U.S. Army Special Forces—the Green Berets—as it was operating as an intelligence organization. Donovan was most fascinated by working with resistance groups within German-occupied territory and conducting daring raids, not by passive intelligence collection. Still, with the help of such operatives as Allen Dulles and Richard Helms he built an espionage service that was to endure beyond the OSS's brief existence to become the CIA's clandestine service. The Research and Analysis section, led by William L. Langer of Harvard

University and including a number of distinguished university professors who answered Donovan's call in 1941–42, also survived as the Directorate of Intelligence in the CIA.

But Donovan was not able to absorb signals intelligence. The navy's code breakers were breaking some Japanese diplomatic communications by 1940 and improved as the war spread into the western Pacific. The army's revitalized code breakers—at Arlington Hall in northern Virginia, where they compiled the now famous Venona decryptions of Soviet agent activities, and with the British at Bletchley Park, where they produced the Ultra intercepts of German communications throughout the war—remained in the War Department. Aerial photography, what would later be known as imagery intelligence, remained primarily within the army air corps, but the OSS had a hand in it.

Nor was Donovan able to absorb all responsibilities and operations in counterintelligence—that is, intelligence designed to uncover hostile intelligence operations against the United States. On the domestic front, the Federal Bureau of Investigation (FBI) had that responsibility but proved no match for Soviet espionage agents and the Communist Party of the United States. Many Soviet spies duped the FBI, the OSS, and every other counterintelligence effort during and after the war. Abroad, the army's Counterintelligence Corps was active in Europe and Japan, where it clashed over turf with the successor to the OSS, the Central Intelligence Agency. Thus the counterintelligence field was weak and increasingly fragmented during and after the war.

As the war came to an end, Donovan again tried to persuade the president, now Harry Truman, to maintain a centralized peacetime intelligence organization above the military departments, directly under the White House. He lost the bureaucratic battle and resigned. Truman abolished the OSS in September 1945. Seeing that the postwar period might not be peaceful, Truman created the

Central Intelligence Group (CIG) in January 1946 as a temporary organization to hold the personnel from the OSS while a review was made of the entire national security apparatus.

The CIG allowed the strong network of OSS veterans to maintain its coherence, but it never gained significant authority or control over the FBI and the military service intelligence chiefs. When the Congress finally decided to pass the National Security Act of 1947, merging the Navy and War Departments into the Defense Department, OSS alumni got the chance to establish themselves permanently in a civilian organization outside of the military. The new law created the Central Intelligence Agency.

The 1947 act also created a director of central intelligence (DCI). This position is distinct from the director of the CIA, but in practice one person has always worn both hats. If there was going to be anything like central management of intelligence, a DCI was necessary because signals intelligence, aerial photographic intelligence, large analysis capabilities, and even larger tactical intelligence capabilities within the military services remained in the newly created Defense Department. If the CIA could not "own" these capabilities, then its director, wearing his DCI hat, could at least manage them. In other words, the CIA was a relatively small part of the larger intelligence services within the armed forces, cocky about its superiority, but insecure because it controlled so few of the resources dedicated to intelligence in the Defense Department and elsewhere.

Double-hatting the director of central intelligence as the director of the CIA limited his ability to stand above and orchestrate the whole intelligence community. Instead, he became the prisoner of the CIA and shared its insecurity about control over other agencies' resources and turf. Periodically, DCIs have risen above those restraints to assert a stronger role in allocating resources efficiently and imposing coordination in place of turf wars, but that pattern has never been institutionalized. In the intelligence community, there is an old saying about the CIA: before it recruits any spies abroad, it must "recruit the DCI," ensuring that he works primarily for its interests.

A few other changes soon followed, most designed to meet particular demands, not part of an overall design of the Intelligence Community. The three military departments in the Defense Department were forced to combine their signals intelligence under the National Security Agency in 1952. In 1960 President Eisenhower created the National Reconnaissance Office, jointly operated by the CIA and the air force, as a research and development organization with great fiscal flexibility to carry out innovative development, and procurement of technical collection capabilities. It was to produce many important technological advances for technical intelligence collection, especially satellite-based systems. Finally, the Defense Intelligence Agency was created within the Defense Department in 1961.

The CIA itself underwent a great deal of change soon after its creation in 1947. Its clandestine service had to be transformed into a peacetime organization and its efforts expanded into many new areas of the world. Langer's Research and Analysis unit became the CIA analysis and production arm. Langer, however, returned to Harvard, leaving Sherman Kent, a professor from Yale, as head of analysis and production. Kent developed a system of national intelligence estimates that continues today. Other branches of the CIA were created to conduct covert operations ranging from propaganda to paramilitary operations, a restoration of the OSS traditions from World War II.

Since that time, the Intelligence Community has remained essentially unchanged in its general outlines. The director of central intelligence had presided over allocation of the resources for all parts of the community, consolidated intelligence needs from all government agencies, and sponsored the production of so-called national intelligence in the form of estimates, the lineal descendants of analysis produced by the Office of National Estimates.

This very brief story of the intelligence restructuring in 1947 and after should make clear (if, indeed, one can ever be "clear" about intelligence!) the stakes in intelligence reform today.

Given the intelligence void existing in 1941, extraordinary measures, including reorganization, were essential. By the end of the war U.S. intelligence was becoming reasonably effective in several areas. Had the OSS been left within the new Department of the Army, it would have soon fallen into neglect. The old War Department had direct access to the White House. Submerged under the new secretary of the army, who is in turn submerged under the secretary of defense, intelligence would have received neither the attention nor the resources it did under the director of central intelligence. Although the CIA failed to anticipate the North Korean attack in 1950, the reorganization yielded major positive results.

But reorganization opened the way to all kinds of new turf wars and bureaucratic struggles by creating a veritable thicket of new intelligence organizations. And it left counterintelligence even more fragmented, with the FBI retaining the dominant but entirely separate domestic counterintelligence role. Thus many of today's problems can be traced back to the reforms under the 1947 National Security Act.

Both military reorganization and intelligence reform were legislated in 1947. Today, intelligence reform is once again being seriously considered at the same time that homeland security reorganization is being designed. In 1947, however, the most prominent intelligence organization was struggling to escape from its cabinet department's control. Today, it and the FBI are struggling to avoid being subsumed under a homeland security department. Although there were good arguments for and against creating an independent CIA, the arguments for it look fully justified in retrospect, notwithstanding the resulting turf wars and misuse of resources. Similarly, good arguments can be made for giving homeland security its own intelligence collection and analysis capability, but when we consider the implications of that approach for the effectiveness of the Intelligence Community, the better arguments are against doing so.

The Intelligence Community itself, as in 1947, desperately needs its own structural reforms. They must be done separately from homeland security reorganization, however, or dysfunctions will be increased, not reduced. At the same time, proper intelligence support to a homeland security department has to be provided. The Intelligence Community, especially a reformed one, can do this. Putting the FBI and CIA into the homeland security department initially sounds sensible, but once we understand how the Intelligence Community works, we see that such a move would create an impossible mess.

A great deal is at stake today, not just in intelligence reform but also in homeland security reform. The clearer our understanding of the tangled issues of intelligence reform, the fewer will be the unintended consequences. And the price of those consequences could be very high.

"Intelligence Reform for Dummies"

A series of computer software instruction manuals with the titles ending in ". . . for Dummies" has become popular among nontechnical computer users, and the concept has now spread to other fields. The arcane language used in the intelligence community can make the uninitiated feel—some would say is *designed* to make the uninitiated feel—like dummies, but the concepts can be grasped by any educated adult. One aim of this book is to cut through the "dummy" factor—the smokescreen of specialized language that keeps outsiders from thinking that they have any business talking about intelligence reform.

Chapter 1 is a more complete version of several of the points made in this introduction. Most careful readers will understand it easily, and to the extent that they do, they will find later sections easier going.

We come to the real substance of reform in Chapter 2. There I methodically clarify intelligence terminology so that people talking about intelligence reform will find it difficult to mislead or talk past each other. Think of it as a Berlitz course in "intel speak." The bane of the confused language problem in the Intelligence Community can hardly be exaggerated.

Take, for example, the term *security*, a word that may seem too simple to require clarification. Its meaning, however, can be distorted to cause inordinate mischief. Some would-be experts insist that security is counterintelligence—if not the whole of it, then a key part of it. From that article of faith, these canon lawyers reason that counterintelligence organizations are responsible for *all* security against foreign intelligence—that is, keeping spies or terrorists out of U.S. organizations and territory. Counterintelligence officials must "protect our secrets," every one of them! To do that, *they* must decide what is secret, and *they* must have the authority to order or direct cabinet officials—the secretary of state and the secretary of defense—to follow *their* security rules. Such rules include the hiring and firing of people with security clearances, the construction and management of buildings, and so on, to prevent hostile intelligence penetrations.

It may surprise intelligence officers, but cabinet secretaries, military commanders, and civil service line managers are not a ventriloquist's wooden dummies; they will not play Charlie McCarthy to the Edgar Bergen of low-ranking (or high-ranking) counterintelligence officers. Line managers from cabinet level downward are normally pleased to receive counterintelligence products, but they reserve the right to act on them as they see fit.

This problem is not just theoretical. In the 1980s, when the newly built American embassy building in Moscow was discovered to be saturated with Soviet listening devices, Secretary of State George Shultz was reluctant to take the full range of security measures needed to neutralize the devices. He was no more enthusiastic about personnel changes and other security measures needed to prevent new penetrations once the building had been cleansed of devices.

At this point the definition of *security* became pivotal. Several officials in the intelligence community who believed that security is the core of counterintelligence insisted that it was their duty to "force State to clean up its act." Even if one fully agrees that Shultz should have been forced to take the penetrations more seriously, it certainly was not—and should not be—within the power of the intelligence community to force him to do so. He worked for the president, not for counterintelligence officials. If he wanted the Soviet KGB to read his mail, that was his choice, not theirs. Only the president, who hired the secretary, could deny him that choice. Only the president was in a position to "force State to clean up its act" over the secretary's objections.

In the event, the president, urged by a couple of other cabinet officers, insisted that Shultz take some actions. He begrudgingly did, but not very effective ones. The most that good counterintelligence can achieve is to show presidents, cabinet officials, and lower-level line managers that they have been penetrated by hostile intelligence. They cannot implement effective security measures against penetrations. But simple definitional disputes, as this case shows, can become the basis for all kinds of bureaucratic wrath and infighting. A detached observer at the time would have been justified in wondering whether Shultz did not prefer to let the Soviet KGB read his mail rather than yield a bureaucratic victory to Secretary of Defense Caspar Weinberger and a covey of counterintelligence officials in his department.

The security dispute is alive and well today among proponents of counterintelligence reform. Several have long argued that security should be the responsibility of the counterintelligence organizations, primarily the FBI. This would give the FBI powers over the president that even the Congress and the Supreme Court do not have! Such proposals reveal a lack of appreciation for basic organizational and management principles, a shortcoming that could produce very messy organizational predicaments in a new homeland security department.

Collection management is another term saturated with definitional problems. It simply means directing the collection of intelligence. At the most general level, it means assigning questions to intelligence collection agencies to answer. Or it can be more assertive, amounting to orders to those agencies. At some point collection management becomes telling which spy to collect what, which satellite to take a picture of what, and so on. It also involves deciding whose requests for intelligence will be served first, whose will be second, and whose may not be answered at all for lack of collection capabilities. It is very much like what editors of newspapers and television news channels do in assigning journalists to stories and deciding which ones are published and on what page in the paper or in which segment of the TV broadcast.

To many senior intelligence analysts, the distinctions are not clear between making requests for collection and "directing" collection when it requires highly technical guidance and instructions. The complexities of collection management are lost on them. Intelligence analysis centers understandably want to control collection management. At some level they should, but does that empower them to hold a list of the CIA's deep cover agents and to know where they are so that these intelligence analysts can manage their collection of information to fit the analysts' demands? Or does it mean that such analysts must be able to direct satellites? Some of them have insisted that it does.

Obviously, the CIA's Directorate of Operations has good reasons not to allow analysts throughout the Intelligence Community to hold lists of its agents. As for directing satellites and other technical collection means, general intelligence analysts are no more capable of doing so than the average reader of this text. Technical collection management of many such systems requires great expertise and experience, but that has not always deterred senior analysts with degrees in the social sciences from insisting that they can handle the task.

The confusion does not stop here. Because intelligence analysis is done in hundreds of places in the military and in civilian agencies, which analysts will get priority in collection management? CIA analysts will insist that they have first priority because they are "central"—they work for the White House. The military commander involved in a battle with Taliban forces in Afghanistan could legitimately dispute that claim. Such cases are not imaginary. Senior officials in intelligence analysis have repeatedly tried to grab full control of collection management, denying the military commander even a hearing for his claim on collection capabilities. Field commanders have on occasion been equally determined that they should have first priority in all cases. Obviously neither the CIA analysts nor a regional military commander is in a position to determine whose demands should be met first from a global viewpoint. It is possible, however, with effectively designed organizational processes and authorities—that is, "doctrine"—for the Intelligence Community to decide, based on senior policy-makers' guidance, whose requests should be met first.

The definitions in chapter 2 may therefore seem elementary, but the implications go well beyond the obvious. In the allocation of people and budgets, for example, small definitional distinctions can mean billion-dollar differences.

Chapter 3 makes it clear that the Central Intelligence Agency is not the same thing as the Intelligence Community. The CIA may be "central," but it is neither the center of the intelligence community nor a large part of it.

I also clarify in that chapter what the director of central intelligence (DCI) really is and what the position could and should be. When members of the intelligence oversight committees in the Congress become frustrated by the "federal," or "confederal," character of the Intelligence Community, they sometimes call for a "tsar," a "director of *national* intelligence" who is "really in charge of all intelligence." In chapter 2 I show that such a position already exists—the DCI—but then explain why the DCI can never be fully in charge of intelligence without creating pandemonium in Defense, State, and several other departments. Were the DCI given full executive authority over the National Security Agency (NSA) and the National Imagery and Mapping Agency, two large intelligence organizations within the Department of Defense, the military services would withdraw their military personnel from both (more than two-thirds of the total in the NSA) and duplicate these organizations within their own ranks.

In short, it would make more sense to put the CIA in the Department of Defense than to take these large intelligence agencies out and put them entirely under the DCI's administration.

The DCI could, however, give up his hat as the director of the CIA and begin to act like the monarch of a federal intelligence community, an unprecedented arrangement that urgently needs to be tried. Until it is, large and long-standing inefficiencies in the use of funds for technical intelligence means cannot be overcome. The

potential savings were in the billions of dollars in the 1980s and
1990s. Today they are probably somewhat less because overall
spending on satellite systems and other technical means has declined.

Chapter 4 deals with intelligence inside the Pentagon. Three
clear messages should be taken from it:

First, almost the entirety of the national intelligence community *can* be brought to bear in support of U.S. military forces in
wartime. Often, however, the intelligence community is not exploited in this way, because the CIA does not like to be subordinated to regional military commanders during a war. A flurry of
publicity was given in the early period of the war in Afghanistan to
the CIA's brilliant performance in providing intelligence about enemy operations. Several months later, however, reports began to
surface about the problems the CIA was causing by operating on its
own in Afghanistan and how it failed as often as it succeeded in
linking up U.S. Green Beret units with Afghan anti-Taliban forces.
The whole story has still not been told, and we cannot be sure
which version is closer to the truth. We do know, however, that during the Persian Gulf War the CIA disappointed and irritated General Schwarzkopf, the theater commander in that conflict.

Second, the military needs two quite different types of intelligence support. Field commanders need timely intelligence about
enemy operations, the kind that the CIA gave U.S. forces in Afghanistan. Over the past two decades, great improvements have
been made in providing this kind of intelligence support, especially
signals and imagery intelligence. Both the Gulf War and the war in
Afghanistan show that trend.

Another kind of intelligence is needed to support the design
of U.S. military weapons and organizations. Aircraft designers
must know in voluminous detail about all foreign air defense systems because U.S. aircraft may have to fly against them. Tank gun

designers must know the details of armor protection on all foreign tanks and other armored vehicles to ensure that U.S. tank guns can defeat them. Designers of U.S. submarines, ships, and aircraft need similar intelligence about foreign weapons that can be used against these naval systems.

This kind of intelligence, to which the CIA contributes much less than do the Defense Intelligence Agency (DIA) and the military services' intelligence organizations, gets almost no public attention. And yet it affects the spending of hundreds of billions of dollars for fielding and equipping U.S. military forces. Contrast the importance of this kind of intelligence with all the publicity given to the Intelligence Community's estimates of Soviet defense spending, which had no effect on any decision making either in Congress or in the Department of Defense. The numbers and quality of Soviet weapons drove Pentagon decisions, not Soviet budgets. Or compare it with the public stir over the CIA's failure to predict the end of the Cold War, which had no influence at all on U.S. policymaking. Can those who complained about this so-called "failure" offer examples of anything the United States could have done better in responding effectively to Gorbachev's decision to end the Cold War? Washington managed the largest strategic realignment in Europe in its history—the reunification of Germany within NATO and the disestablishment of the Warsaw Pact—without a war. Important intelligence is usually boring and not newsworthy, while newsworthy intelligence is either unimportant or destroys important sources (agents and communications).

The third message about Pentagon intelligence sources concerns so-called tactical intelligence. The money for it—more than $10 billion annually—is not and cannot be in the intelligence community budget. The director of central intelligence—or a new director of national intelligence—cannot possibly manage army, navy, air force, and marine intelligence needs at the tactical level.

Thus a large part of the overall spending on intelligence will never be in the "national" intelligence budget.

At the same time, large national capabilities can collect and disseminate intelligence to tactical forces, often within minutes after it has been collected and processed. This practice is far advanced in signals intelligence, but the CIA's turf interests have prevented equal progress in imagery intelligence, and they have virtually precluded progress in clandestine human intelligence.

Chapters 5, 6, and 7 are about the major intelligence collection disciplines—signals, imagery, and human intelligence. These are radically different methods of collection. They require equally different collection technologies. They have different organizational cultures. They are truly different worlds. In caricature, signals intelligence is done by mathematical geniuses and nerds, imagery intelligence is the business of photography buffs, and human intelligence is the world of James Bond and con artists. These, of course, are gross distortions, but they do capture something of the differences in cultures.

A major thesis of this book is that each of the three collection disciplines needs to have a national manager responsible to the director of central intelligence. If these managers, regardless of the agency in which each is located, have control of all the resources (money and people) in their disciplines, then they can be held accountable by the DCI for getting more intelligence "output" for every dollar of "input." Today, there is no way for the DCI to hold anyone accountable for input-output relations in collection activities. With a budget of perhaps $30 billion, sustaining the current arrangements borders on the criminal. Creating national managers for each intelligence discipline with full power over resources has the potential to save billions of dollars. No other reform measure can generate such large gains in efficiency.

Instead of considering a system of national managers, most

ostensibly expert discussion of reforms in this area deals with non-issues. A lot of silliness is talked about trade-offs among these three collection disciplines. "If we just had more human intelligence and fewer satellites," the argument might go, "we would do better against terrorists," or against drug traffickers, or some other target. Few people within the intelligence community have a sufficient picture of what all three disciplines actually collect to speak knowledgably about such trade-offs; those who have such a picture do not believe that thinking in terms of trade-offs is useful. If a car owner were asked to trade off his engine against his tires and steering wheel, he would stubbornly reject the demand. He needs all three. He might buy cheaper tires, or a smaller engine, but he could not give up either and still drive his car.

Another silly debate that occupies intelligence experts involves "raw" intelligence versus "all-source" intelligence—that is, intelligence analysis based on all available collection sources, from espionage reports and communications intercepts to photography and such unclassified materials as media reporting and scholarly books and articles. The Directorate of Intelligence at the CIA and some parts of the Defense Intelligence Agency insist that all-source intelligence alone must be supplied to users, never raw signals intercepts, or imagery reports, or clandestine agent reports. As "non-professionals," the users are apt to draw mistaken conclusions. So reason the wise men of all-source analysis.

Certainly, it helps to have intelligence from all three disciplines and unclassified sources on any single question, and analysts who know a target well can probably draw sounder conclusions from the raw reports. But what if the user of the intelligence is a battalion commander engaged in combat in Afghanistan who wants an imagery report of what is beyond the hill in front of him? Or an air operations center during the Persian Gulf War that wants instant knowledge of Iraqi air defense activity? Or a diplomat negotiating a treaty in Geneva who wants to know what his negotiating partner's president just said in a speech? Or a national security adviser helping the president decide how to deal with a crisis requiring an immediate decision, such as approving orders to allow U.S. aircraft to shoot down Libyan fighters? Should all of these users be forced to wait until the CIA or the DIA or the Intelligence and Analysis Bureau at State reviews the raw reports and anoints them with the holy water of an all-source analysis? In every one of these illustrative cases, the local user is in a better position to make good sense of the raw intelligence than is an all-source analysis center far from the action. Remote all-source analysis centers seldom have a genuine sense of the urgency and immediate needs of a military commander, or a diplomat, or the president. Moreover, so-called raw intelligence reports are not so raw. Because they have been processed, they have some stand-alone character. In military operations, signals and imagery reports can reach a user within five or ten minutes after the occurrence of an observed event. And combat responses are made just as quickly. Routing such raw intelligence through the all-source process at the CIA, DIA, or some other center, often merely ensures that the intelligence will arrive too late to be of use.

The solution to this nonsense argument about raw versus all-source intelligence is to have both. To do so requires a second model of intelligence analysis in addition to the central processing model demanded by the proponents of all-source exclusivity. The distributed processing model, which is discussed at length in chapter 3, makes it possible for users to receive tailored analysis done close to the decision maker by intelligence analysts familiar with the user's unique needs. To call this a distributed processing model is to draw an analogy with the shift in computers from large mainframe systems with many users to personal computers with microprocessors located wherever individual users chose and capable of operating autonomously. This analogy is especially useful later in the book when I deal with issues of who should produce finished intelligence analysis.

The raw versus all-source intelligence issue is at the heart of deciding how best to provide intelligence support to a homeland defense department. Giving the responsibility to the CIA or the FBI makes no sense at all. If the new department is to have first-rate and timely intelligence, it must have many widely dispersed analytic centers, each supporting one of the various agencies and activities within the department. Each of these analytic centers needs to be able to "subscribe" to the national collection systems—signals intelligence, imagery intelligence, human intelligence, and counterintelligence. They will also want to build files, central databases, and other such resources within the department, some centrally located but locally available by internet, some dispersed to provide ready access to users. Putting the FBI or the CIA or both in the homeland security department would break them away from the larger Intelligence Community, with its vastly superior collection capabilities. It would defeat the very purpose the homeland security department was meant to serve.

Chapter 8 deals with counterintelligence. Counterintelligence, again, is simply information about an adversary's intelligence operations, capabilities, agents, collection technology, and so on. It is *not* security. It is intelligence on which security policies should be based. Nor is it intelligence about an adversary's policymaking or military operations or other nonintelligence capabilities and activities.

The key message in chapter 8 is that criminal law enforcement organizations have never been effective at counterintelligence, and they never will be. Yet, the counterintelligence agencies of three of the five organizations that have such operational agencies (the FBI, the Naval Investigative Service, and the air force's Office of Special

Investigations) are primarily criminal investigative agencies. Counterintelligence is a secondary mission for them. Counterintelligence is organizationally separate in the CIA and in the army.

The reasons for the counterintelligence ineffectiveness of law enforcement agencies are not so complicated. The two tasks require fundamentally different cultures. Law enforcement agencies thrive on media attention. Counterintelligence agencies perish from it. Criminal catchers are hopeless at catching spies because spies are generally much smarter and more patient than criminals, and they enjoy the support of a foreign government. Counterintelligence agencies need not have arrest authority to uncover spies. They can leave it to law enforcement agencies to make arrests. The technology and sophistication of espionage transcend the characteristics of most criminal activities, allowing spies to overwhelm cops.

The United States has always had miserable counterintelligence capabilities. Spies like Robert Hanssen and Aldridge Ames are not exceptional. Spies were placed in top posts in several departments of the U.S. government in the heyday of J. Edgar Hoover's FBI. Hoover proved a pushover for Soviet intelligence operatives, a mere amateur whose bureau was swindled, misled, and thoroughly thrashed in the espionage war between the United States and the Soviet Union. There are and have been outstanding counterintelligence officers in the FBI, but they have never been properly supported. Their culture can never dominate an agency that is three-fourths cops, one-fourth spy-catchers.

For all its flaws, the Intelligence Community has no peer in the world when all of its capabilities are taken together. Hostile intelligence services have dealt it blows in some areas, but none—not even the Soviet KGB and its military partner, the GRU, during the Cold War—has the full array of capabilities that exists in the U.S.

Intelligence Community. At the same time, it has weaknesses that cry out for basic changes, not just policy tinkering and pep talks about doing better. The burden of the remainder of the book is to show why this is true and what needs to be done to change things for the better.

Glossary

Note: Except where indicated, all terms are defined from the perspective of the United States.

Air Force Intelligence Agency (AIA). The air force's signals intelligence organization, which serves as its "service cryptologic element" for the National Security Agency.

Assistant chief of staff for intelligence (ACS/I). The primary intelligence staff officer for the Department of the Air Force.

Assistant secretary of defense for command, control, communications, and intelligence (ASD/C3I). An assistant secretary, who, in addition to other duties, is the secretary of defense's staff aide overseeing intelligence resource management issues in the Pentagon.

Bureau of Intelligence and Research (INR). The intelligence analysis and research organization within the State Department.

Central Intelligence Agency (CIA). An autonomous intelligence agency created by the National Security Act of 1947.

Chief of naval operations (CNO). An admiral and senior naval officer in the Department of the Navy who is also a member of the Joint Chiefs of Staff.

Chief of staff of the air force (CSAF). A general and senior air force officer in the Department of the Air Force who is also a member of the Joint Chiefs of Staff.

Chief of station (COS). The head of a CIA clandestine staff normally located in a U.S. embassy abroad.

Communications security (COMSEC). All means of securing communications from hostile intelligence interception, especially cryptologic devices for encoding transmissions.

Community Management Staff (CMS). The staff that assists the Director of Central Intelligence in management of the Intelligence Community.

Consolidated Cryptologic Program (CCP). The national-level program budget for signals intelligence.

Counterintelligence (CI). Any intelligence about the capabilities and operations of foreign intelligence services working against the United States.

Counterintelligence Corps (CIC). The army's counterintelligence organization created during World War II and disestablished in the early 1960s.

Covert action (CA). Any effort by the U.S. government to influence another country's policy in ways such that the United States is not seen as responsible for the effort. This includes but is not limited to paramilitary operations.

Criminal Investigation Division (CID). The army's law enforcement agency.

Cryptologic Support Group (CSG). Teams of specialists that the National Security Agency provides to military commands to assist in their receiving signals intelligence support.

Defense Cryptologic Program (DCP). The Pentagon's budget for purchasing cryptologic means for encoding and securing communications.

Defense HUMINT [human intelligence] Service (DHS). The military clandestine service in the Pentagon.

Defense Intelligence Agency (DIA). The defense department's all-source intelligence-analysis organization and manager of several other Pentagon intelligence activities.

Defense Investigative Service (DIS). The Pentagon agency that makes personnel background checks and monitors security measures for the Defense Department.

Deputy chief of staff of the army for intelligence (DCSINT). The primary intelligence staff officer in the Department of the Army.

Deputy director for operations (DDO). Head of the CIA's clandestine service.

Deputy director of central intelligence (DDCI). The DCI's deputy.

Directorate of Intelligence (CIA/DI). The CIA's intelligence analysis and production organization.

Directorate of Operations (CIA/DO). The CIA's clandestine service.

Directorate of Science and Technology (CIA/S&T). One of the three major directorates in the CIA. Its major task is managing the CIA's part of the National Reconnaissance Office and its program activities.

Director of central intelligence (DCI). The position created by the 1947 National Security Act to head the U.S. intelligence community and serve as the president's intelligence officer.

Director of military intelligence (DMI). The primary intelligence staff officer in the U.S. Marine Corps.

Director of naval intelligence (DNI). The primary staff intelligence officer in the Department of the Navy.

Federal Bureau of Investigation (FBI). The federal law enforcement agency in the department of justice which also has a national security division devoted to counterintelligence.

Fiscal year (FY). The budget year, beginning 1 October and ending 30 September the following year, for the federal government and the Congress.

Foreign Broadcast Information Service (FBIS). An organization sponsored by the CIA that monitors a wide range of foreign radio and television broadcasts as well as foreign newspapers and journals, translating many of the broadcasts and articles into English.

General counsel (GC). The senior legal aide serving the director of the Central Intelligence Agency. The other major intelligence organizations, such as the National Security Agency, also have a general counsel.

General Defense Intelligence Program (GDIP). Those parts of the program budget for intelligence in the Pentagon under the management of the director of the Defense Intelligence Agency (DIA). It includes monies for parts of each of the military services' intelligence programs as well as for joint intelligence programs under the DIA.

G-2. The primary staff intelligence officer on a division or corps staff in the army.

Human intelligence (HUMINT). Intelligence collected by human sources rather than primarily technical means. It includes both secret and unclassified collection activities.

Imagery intelligence (IMINT). Intelligence collected through photography and all other types of image-making technologies.

Information security (INFOSEC). Expansion of communications security to include broader security concerns involving computer-based information and the internet.

Information warfare (IW). In its narrow definition, it means attacking an enemy's communications and computer systems. In its broad definition, it can include all kinds of information use, such as propaganda and policy actions designed to mislead, frustrate, or otherwise confuse or degrade the enemy's overall information picture of a conflict.

Inspector general (IG). The senior official serving the director of central intelligence who is responsible for periodic inspection of the

CIA and all its subunits to evaluate adherence to all rules, regulations, and laws. Inspectors general also have positions in the other major intelligence organizations in the Intelligence Community.

Intelligence and Research (INR). *See* Bureau of Intelligence and Research (INR).

Intelligence and Security Command (INSCOM). The army's worldwide command that manages logistics, finances, and many other support activities for army intelligence organizations above division and corps level.

Intelligence Community (IC). The name used to identify all of those intelligence agencies and activities which come under the tasking authority and the program budget authority of the director of central intelligence.

Intelligence Community Executive Committee (IC/EXCOM). A committee of most of the heads of agencies within the Intelligence Community, chaired by the director of central intelligence, which deals with resource management and administrative policy issues.

Intelligence Support Activity (ISA). A small intelligence organization created to provide tactical intelligence in support of the Iran hostage rescue mission in 1980.

Joint Chiefs of Staff (JCS). The committee of the chiefs of all the military services that advises the secretary of defense and the president on military affairs.

Joint Military Intelligence Program (JMIP). One of several program budgets in the Pentagon designed to coordinate intelligence development and procurement programs for tactical intelligence systems.

J-2. The primary staff intelligence officer on any joint staff.

Measurement and Signature Intelligence (MASINT). Technical intelligence about adversaries' weapons and technical systems needed to support the development and targeting of U.S. "smart" weapons systems.

Military Intelligence Board (MIB). Chaired by the director of the Defense Intelligence Agency (DIA), it consists of the four military service intelligence chiefs: the deputy chief of staff of the army for intelligence, the air force assistant chief of staff for intelligence, the director of naval intelligence, and the marine director of military intelligence. It may also include one or two deputy directors of the DIA.

National Clandestine Service (NCS). At present not an official term, it is proposed as a new name for the CIA's Directorate of Operations to emphasize its operational control over any other clandestine capabilities that exist elsewhere in the Intelligence Community.

National Counterintelligence Service (NCIS). An organization proposed in this book to manage all counterintelligence activities.

National Foreign Intelligence Board (NFIB). A board consisting of the most senior intelligence officials in the Intelligence Community, chaired by the director of central intelligence (DCI), which reviews and approves national intelligence estimates, other national intelligence products, and other matters concerning intelligence analysis and production that the DCI chooses to put before it.

National Foreign Intelligence Council (NFIC). An earlier name for the Intelligence Community Executive Committee.

National Foreign Intelligence Program (NFIP). The combined program budgets of all the agencies in the Intelligence Community, managed and approved by the director of central intelligence.

National Imagery and Mapping Agency (NIMA). Created in 1997, it is responsible for all national-level imagery intelligence production and production of maps, primarily for the military services' use in operations and planning.

National Imagery and Mapping Program (NIMAP). The program budget for imagery intelligence and map production within the National Foreign Intelligence Program.

National Intelligence Council (NIC). Composed of the national intelligence officers and subordinate directly to the director of central intelligence, it manages the production of national intelligence estimates and other national intelligence products.

National intelligence estimate (NIE). An intelligence assessment addressing either a broad or narrow topic that reflects the combined judgments of all intelligence agencies on the National Foreign Intelligence Board, which then approved it.

National intelligence officer (NIO). A member of the National Intelligence Council who is responsible to the director of central intelligence for managing national intelligence production in a specific area or functional specialty.

National Photographic Interpretation Center (NPIC). An analysis section of the National Imagery and Mapping Agency.

National Reconnaissance Office (NRO). A joint air force–CIA research, development, and procurement agency.

National Reconnaissance Program (NRP). The program budget for the National Reconnaissance Office, part of the National Foreign Intelligence Program.

National Security Agency (NSA). The agency charged with signals intelligence and information security.

National Security Division (NSD). The division within the FBI responsible for counterintelligence.

Naval Investigative Service (NIS). The law enforcement and counterintelligence organization in the Department of the Navy.

Naval Security Group (NSG). The navy's signals intelligence unit, also known as the navy's service cryptologic element.

Office of Energy Intelligence (OEI). The top intelligence staff element in the Department of Energy.

Office of Intelligence Support (OIS). The top intelligence staff element in the Department of Treasury.

Office of Naval Intelligence (ONI). The staff office in the Department of the Navy responsible for intelligence support.

Office of Special Investigations (OSI). The navy's criminal investigation organization, which also has counterintelligence responsibilities.

Office of Strategic Services (OSS). An intelligence and special operations organization created in World War II under the Joint Chiefs. Although dissolved in 1945, most of its personnel remained to become the core of the CIA when it was created in 1947.

Office of the Secretary of Defense (OSD). The staff sections that work directly for the secretary of defense, including the deputy secretary of defense, the undersecretaries, assistant secretaries, and others.

Operational control (OPCON). A military term for arrangements for control over units by joint commanders without full command for all other purposes, such as feeding, clothing, discipline, pay, housing, equipment, and so on.

Office of Program Analysis and Evaluation (PA&E). A staff section in the Office of the Secretary of Defense that manages program budgeting for the secretary and does analysis to support the secretary's budget decision making.

Planning, programming, and budgeting system (PPBS). A system introduced in the Pentagon in the 1960s that groups budget items together in support of defense missions. It clarifies the connection between inputs of dollars and outputs of combat capabilities.

Program Analysis and Evaluation (PA&E). *See* Office of Program Analysis and Evaluation (PA&E).

Remotely piloted vehicle (RPV). An aircraft without a pilot, controlled from the ground, and often used to carry imaging systems to acquire intelligence from the air.

Research, development, testing, and evaluation (RDT&E). The term used to describe defense spending on the full process of creating, developing, and testing military weapons and equipment.

Resource management (RM). All those activities related to deciding what resources (people and money) are needed, making the case to the Congress, then using those resources appropriated by Congress.

Service cryptologic element (SCE). The technical name for each military service's component of the National Security Agency's operational capabilities: the Naval Security Group, the army's Intelligence and Security Command, and the Air Force Intelligence Agency.

Side-looking airborne radar (SLAR). A radar imaging system carried in an aircraft.

Signals intelligence (SIGINT). Intelligence derived from intercepted electronic communications.

Support to military operations (SMO). Provision of intelligence to the field commands worldwide during peace and war.

Tactical exploitation of national capabilities (TENCAP). A National Reconnaissance Office program funded by the military services, not the director of central intelligence's National Foreign Intelligence Program.

Tactical reconnaissance and related activities (TIARA). The collection of military service programs devoted to tactical intelligence capabilities. It is not part of the director of central intelligence's National Foreign Intelligence Program.

Unmanned aerial vehicle (UAV). Another term for a remotely piloted vehicle.

U.S. Special Operations Command (USSOCOM). The unified command that employs Rangers, Seals, Special Forces units, and several other special operations capabilities.

Why Intelligence Reform?

Studies on intelligence reform have become a cottage industry. Why do we need another one? The answer is simple: most of them have attempted to doctor the symptoms, not the illness. The skeptical reader will object that this claim is not self-evident. But consider the public record.

Since Congress began investigating the Intelligence Community in the mid-1970s, the issue of intelligence reform has been raised repeatedly. During the Carter administration several initiatives were taken to implement some of the ideas produced by a committee chaired by Senator Frank Church (D-Idaho), but no fundamental structural change occurred. Several times in the 1980s congressional oversight committees raised the reform issue, and Senator David Boren (D-Okla.) actually drafted legislation for several structural changes in the Intelligence Community. The House committee offered an alternative draft, but neither bill became law. In 1994 Senator John Warner (R-Va.) introduced legislation for a presidential commission to consider Intelligence Community reforms, and his bill became law. The resulting commission produced its report in early 1996.[1] At the same time, several unofficial intelligence reform studies produced a flurry of activity and a wide variety of proposals.[2]

Most of these reform efforts were inspired by sensational problems and episodes within the Intelligence Community, espe-

1

cially in the Central Intelligence Agency (CIA). The Church committee was outraged by evidence that the CIA had attempted assassinations as part of covert actions in the past, and that the army's counterintelligence units had been used to help the Federal Bureau of Investigation (FBI) keep track of antiwar movement leaders in the late 1960s and early 1970s. In the late 1980s, CIA covert actions in Central America and related activities by National Security Council staff members became the focus of renewed interest in intelligence reform. In the 1990s a number of incidents within the CIA, some having to do with personnel policies, others involving serious operational failures, and still others involving National Reconnaissance Office accountability for funds, brought the issue once again to congressional and public attention. The revelation in 1994 that Aldrich Ames of CIA's Directorate of Operations was a KGB agent produced a new and unprecedented level of concern.

After almost three decades of such episodes, no fundamental reform has occurred. Virtually all congressional investigations and reform studies have merely focused on the scandals and raised policy issues. For example, should the CIA be permitted to carry out assassinations? Should Army Counterintelligence be involved in domestic surveillance of civilians? Should CIA clandestine officers be allowed to have cover as journalists? Should the Intelligence Community budget be made public? And so on.

Almost none of the congressional committees' efforts at reform have addressed structural, organizational, and management issues. The major exception was Senator Boren's draft legislation that directed certain organizational changes, notably the creation of a national agency specializing in intelligence gleaned from photography, radar and more sophisticated imaging technologies imagery. A second partial exception was the 1996 organizational redesign proposal by the House Permanent Select Committee on Intelligence.[3] A presidential commission of 1996 was instructed to examine the roles and missions of all the parts of the Intelligence Community—an invitation to deal with the structural issues—but it effectively declined the invitation.

One major feature of the Boren bill of the late 1980s, the creation of a national imagery agency, was actually accepted, and the National Imagery and Mapping Agency was created on 1 October 1996. A few additional but minor organizational steps have been taken in connection with the support staff for the director of central intelligence (DCI), formerly called the Intelligence Community Staff, now reorganized and renamed the Community Management Staff. Immediately after the Ames case broke, a flurry of activity centered on the traditional FBI-CIA dispute over counterintelligence turf, but interest soon abated.

With few exceptions, however, structural reform has been largely ignored. This is both strange and unfortunate. Several senators and House members proved reluctant to delve into the structural issues. The CIA and most directors of central intelligence have also resisted serious review of the structural problems. Thus a quiet and informal consensus that nothing structurally is wrong has prevailed, not only in Congress and in the Intelligence Community itself but in the presidential commission as well. Private-sector intelligence reform studies have followed suit.

The reasons why are not entirely clear, especially after Senator Boren opened the question of deeper problems in need of structural solutions. Senator Dennis DeConcini (D-Ariz.) complained that the problem was the "culture" at the CIA. Organizational cultures are normally the products of structural conditions. The accumulation of dangerously embarrassing incidents should have suggested to both administration officials and the congressional committees that they were looking at the symptoms, not the underlying ills.

Do serious structural and organizational problems really exist? One does not need access to classified information to answer yes. It would be extraordinarily surprising if there were not. That Aldrich Ames of the CIA and Robert Hanssen of the FBI could be recruited by the KGB, meet KGB case officers undetected for years, and make "dead drops" in Washington, D.C., under the nose of the FBI suggests that more than policy problems are involved. Little known to the public but notorious in military circles, failures in CIA intelligence contributed to the failed attempt in 1980 to rescue hostages in Iran and compromised operations in the Persian Gulf War. Other intelligence failures, as well as bureaucratic turf fights between the National Reconnaissance Office and other agencies, are evidence of deeper problems. Regardless of the individuals in charge, the structural arrangements and the problems they create have remained constant or have worsened.

Since the Intelligence Community structure settled into place some forty years ago, it has remained essentially unchanged. Meanwhile, technologies have advanced at a blinding pace over those decades. In the 1960s transoceanic communications were fairly limited. As a result, many technical intelligence activities had to be located in Europe and East Asia. A decade later, many of those activities had been moved to the United States, as space-based communication capabilities made possible radically different and more effective ways of gathering, processing and producing, and distributing intelligence. In the quarter-century since, communication has advanced exponentially in speed and sophistication. The communications revolution alone provides grounds for suspecting that major structural reforms in the Intelligence Community are long overdue. Some adaptations have been made, but not enough to allow full exploitation of the technologies.

The introduction of intelligence collection systems in space, for example, created radically new possibilities, and many of them were exploited in the 1960s and 1970s. As time passed, however, and as the number and variety of space systems increased, new ways to exploit them became possible that could not have been foreseen. Most such opportunities, however, require organizational changes, and very few were made. A national imagery agency should have been created at least twenty years ago, but bureaucratic turf concerns kept the very idea from being considered. Nor does the belated creation in 1996 of the National Imagery and Mapping Agency assure that it will be given all the authority required to make the most of imaging systems.

Structural problems afflict not only intelligence collection but also intelligence analysis. During the Cold War this was conspicuous in the continuing debate between the CIA and the Defense Intelligence Agency (DIA) on estimates of Soviet military capabilities. The gross underestimate of Soviet military expenditures can be explained largely as the result of competition that caused each of these agencies to be less concerned with the truth of matters in the Soviet Union than with proving the other wrong in the eyes of the Congress. The needs of executive branch policymakers, for whom this intelligence was primarily produced, tended to be secondary in both agencies' calculations. The handling of intelligence analysis on al Qaeda in the weeks and days leading up to the events of 11 September 2001 offers another example of problems in both analysis and distribution of intelligence. The FBI's jealousy over its counterintelligence turf, not only vis-à-vis the CIA but also vis-à-vis the military services' counterintelligence analysis (as well as operations), is a similar symptom of structural problems.

These examples suggest that structural problems distorted U.S. perceptions of Soviet military and economic capabilities and permitted the KGB to penetrate our own intelligence services to a degree that was preventable.

Failure to make appropriate organizational changes to exploit evolving technology has also wasted money. The costs of technical intelligence collection systems dwarf the costs of intelligence analysis and clandestine human intelligence operations. Yet efficiency has rarely been considered in the purchase of technical systems. In the 1950s and 1960s, there was not yet adequate experience with most new technologies to allow meaningful efficiency comparisons. In some cases it could also be argued that Intelligence Community technologies more than paid for themselves as they migrated into the civilian economy. For example, development of modern digital computational means—computers—occurred almost entirely as a result of the National Security Agency's research and development efforts in the 1950s. IBM and CDC essentially got their start in modern computers from National Security Agency funding, and without it, we might be two decades behind where we are in computers today.

By the 1970s and 1980s, however, these arguments for ignoring the burgeoning costs of technical collection programs were as outdated as the displaced technologies. But bureaucratic processes and organizational interests stood in the way of examining costs and reducing them. Predictably, inefficiencies mounted. Covered by secrecy, arcane organizational practices, and technical complexities, these accumulating structural-management issues have remained largely opaque to the congressional oversight committees and also to high-level officials in the executive branch. Neither DCIs nor secretaries of defense have really understood them fully.

None of the accumulating inefficiencies and missed opportunities for better and more efficient exploitation should be terribly surprising in organizations dealing with fairly rapid change in leading-edge technologies. IBM, AT&T, GMC, Chrysler, and other large industrial firms have been slow to make fundamental structural changes, too, but they did make them, in the face of serious financial difficulties resulting from business competition. Management at all levels in organizations tends to resist change, especially structural reform. But if change is necessary for AT&T, is it not also necessary for the Intelligence Community?

Critical scrutiny of the Intelligence Community is clearly justified, scrutiny of its structure, organization, and management arrangements. I am *not* concerned here with issues that received a lot of media attention in the 1990s, like whether or not the CIA should put agents under cover as journalists. I am not concerned with whether or not the CIA predicted the collapse of the Soviet Union. I am not concerned with proper security rules and techniques for preventing another Ames case. I am not concerned with whether more intelligence attention should be put on the Third World, terrorism, nuclear proliferation, Russia, or economic affairs. Nor am I concerned with whether the United States should engage in covert actions, or whether the United States should have an Intelligence Community. (Those last questions, however, are relevant to the structural arrangements for intelligence support to a new Department of Homeland Security.) These are issues treated elsewhere, and all of them are policy problems. They are important, but they are different from management and structural problems. Thus they will be treated here only when they appear as symptoms of structural problems. As the events of 11 September have once again forced intelligence issues onto the agenda, the initial reactions have been to treat them primarily as policy matters, ignoring the underlying structural issues.

To clarify this distinction with a metaphor, assume that the Intelligence Community is a ship which has had numerous problems on its recent voyages. It is now in harbor for repairs. Most of the ship's investors ask where it should sail next, when it should sail, what flag it should fly, what color to paint the ship. Here those questions will be set aside and primary attention focused on the ship's hull, its engines, its navigation gear; I will look for malfunctions and prescribe needed repairs. When the repairs are completed, it can then sail anywhere, any time, under any flag, with any passengers. These choices are up to executive branch policymakers and military officials served by the Intelligence Community.

2

Essential Dogma and Useful Buzzwords

The major problem confronting all discussions about reform of the U.S. Intelligence Community is the absence of a commonly understood and accepted doctrine—a single set of terms, rules, and practices—for intelligence organization, operations, and management. Virtually every agency within the Intelligence Community has either a unique doctrine or none. The director of central intelligence has never promulgated a unifying set of principles. The Community Management Staff holds some vague ideas about the topic, but these have never been fully developed or promulgated as official guidance. The Central Intelligence Agency (CIA), the National Security Agency (NSA), and the Defense Intelligence Agency (DIA) have internal standard procedures but no clearly articulated doctrine. Between agencies, the procedures are generally incompatible. No formal doctrine governs the intelligence staff structure within the military's joint system of operational commands—forces comprising units from various commands of all three military services—although a few standard concepts tend to prevail in several staff sections. The military services vary in their conceptions of the intelligence process and functions. The army has the most rigorously articulated intelligence doctrine; the navy's practice is fairly well standardized but less formal than the army's; and air force practices differ significantly among major commands.

Even less articulation of standard practices prevails within civilian cabinet departments and agencies for their intelligence components, which are typically very small staffs working primarily on intelligence analysis.

As I have emphasized, vast technological advances have radically changed the ways information is collected and communicated during the half-century of the Intelligence Community's development. The evolution of global communications networks with vast bandwidth for transmission brought a structural revolution in many aspects of military intelligence and intelligence support to diplomacy.

For example, most of the order-of-battle intelligence on Soviet forces in Eastern Europe was developed and maintained within the European Command as late as the mid-1960s. Space-based collection and transoceanic high-speed communications soon thereafter allowed this activity to be centralized in various agencies in Washington, D.C. But little effort was made to generate a systemic doctrine that could make sense of these changes to the agencies they affected and illuminate how they would be adapted to crisis operations and wartime. Accordingly, U.S. forces in Europe did not always benefit from the centralization process. Parochial bureaucratic battles followed. Some commands in Europe, especially in the army, felt cut off, deprived by what they perceived as the "national systems" that were not responsive to their intelligence needs. Many air force units had a similar reaction. As a result, army and air force commands tried to develop independent approaches, especially for wartime operations, because they either did not realize that national systems could be made to support tactical operations in Europe or they did not trust that such systems would give them priority support in an emergency.

At the national level within the Intelligence Community, little genuine effort was made to rectify the situation in Europe, or in the

Pacific Theater, where similar changes occurred. In part this was due to the primary concern of the CIA and the National Security Agency with supplying intelligence to the user community in Washington, but it was also due to simple ignorance at the national level both about the needs of military forces and of the ways that communications could be developed to meet them. Much lip service was given to "support for military operations," but little of note was done until the early 1980s, and even then it was done on an ad hoc basis almost entirely by the National Security Agency and the Defense Intelligence Agency.

Within the Washington Intelligence Community, numerous efforts were made by the DCI's Intelligence Community Staff, now the Community Management Staff, to come to grips with these problems. This staff actually created a system for formal intelligence requirements and tried to tie resource allocations to those requirements. Still, progress was slow, and bureaucratic struggles were intense.

In the 1980s a standing committee reporting to the DCI attempted repeatedly to relate budget inputs to intelligence production outputs according to perceived user demands. But almost none of the members of the committee knew how to assess the connection between inputs and outputs so that the DCI could make effective program decisions in cutting, shifting, or increasing resources. The committee included the heads of all the major intelligence activities, but scarcely anyone with a working knowledge of these input-output connections. In most cases, *such people did not exist,* because the Intelligence Community had grown up without a resource-management doctrine. In fact, the CIA has from its very inception striven to prevent such arrangements because its leaders believe that its influence—"leverage"—over the Pentagon is enhanced by fragmented resource management responsibilities.

The arcane and insular character of most intelligence activities also contributed to a lack of understanding of the problems confronting the Intelligence Community. Senior officials most often spend careers in a single agency, and therefore they arrive in senior posts largely ignorant of other parts of the Intelligence Community. Without a doctrine for intelligence operations and management, they have behaved as any organization theorist would predict: they have defended the parochial resource interests of their home agencies. This kind of behavior has consistently reinforced the split in the Intelligence Community between the national agencies (CIA, NSA, DIA) and military tactical commands, and it has produced long-standing fragmentation among the national-level agencies. This is not to say that the Intelligence Community has made no progress; rather, it is to point out that over time it accumulated many structural obstacles and inefficiencies that increasingly slowed progress. Such hardening of the institutional arteries was inevitable in the absence of clear principles for judging the effectiveness and rationality of changes. In a word, there is very little corrective feedback from output performance to the resource allocation process.

Until a set of principles and concepts is specified and made official doctrine, this dysfunctional behavior will continue. Within the Intelligence Community, a Tower of Babel long ago replaced a language of mutual understanding. Only occasional crises, usually military conflicts, have forced a degree of improvisation that has overcome the more conspicuous dysfunctions, but often such improvement has been temporary. That such improvisation has often proved effective demonstrates that the problems of coordination are solvable. That improvisation is necessary in every case demonstrates that an established doctrine would eliminate the necessity for serial reinvention of the wheel.

In this chapter I will therefore spell out an Intelligence Community doctrine. It need not be the last word, but it will serve as the starting point and the guidance for all of my further analyses and resulting recommendations. It draws on two sources for concepts and principles of intelligence operations.

First, it takes the army's basic pattern emerging from World War II and in the immediate postwar decades. The navy and the air force patterns are too specialized and particularized to serve as a general model, although their intelligence operations can be adapted and understood in the context of the army's doctrine. Their operations are heavily based on a few large weapons systems while the army's operations are far more diverse and complex, demanding a more generalized approach.

Second, intelligence operations have much in common with news operations—the press and television. Because the media have adapted so successfully to modern technology, they provide another model from which certain principles can be drawn.

For management of resources, the models from which principles are taken here include general organization theory, business firms, and nonprofit organizations. Business principles alone are not adequate because the Intelligence Community does not sell its products in a competitive market. Thus the nonprofit case must also be included—not just private-sector organizations but also public agencies, in particular the Defense Department, which has adapted the Planning, Program, Budgeting System for managing resources.

Specification of Functions

Intelligence operations can be divided into a finite set of functions. The ones listed here are not innovative but rather long-standing and generally accepted throughout the Intelligence Community. They are a) collection management (including technical collection management), b) collection, c) analysis and production, and

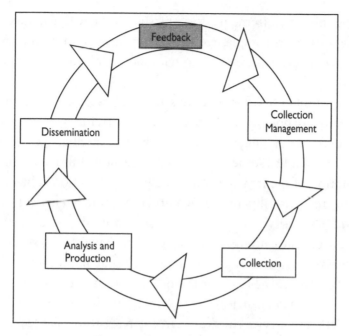

Figure 1. The Intelligence Cycle

d) dissemination. Intelligence operations, like news operations, have a cycle (figure 1). It starts with collection management. The collection manager must determine what intelligence is needed, just as editors decide what issues to cover with reporting. Next the collection manager must issue directions to collectors, telling them what information is needed, just as editors dispatch reporters to get stories. The collectors report back to the collection manager, who then gives the information to analysts. Analysts develop that information—analyze it—into intelligence products. In news agencies, this is the editorial part of the cycle. The products must then be disseminated to the user who stated the initial requirement, and to any other user who needs them. The users provide feedback to the collection manager, expressing satisfaction or discontent and conveying new requirements for answers to new questions. This process, of course, is analogous to the printing and publication of newspapers or magazines, or to television and radio scripting and broadcast.

A number of other activities are often associated with intelligence but are not part of the process. Security, for example, is often considered an intelligence function by the Intelligence Community. The DCI has some security policy responsibilities, but implementation of security is *not* an intelligence function. That is, it is not the responsibility of the DCI and other intelligence officials to provide "security" to intelligence users, military, or civilians, although occasionally they are believed to have that task, or they themselves believe they have it. The result is debilitating turf fights. Security can only be a "command" function (or "management" function in civil agencies). The secretary of defense, unified commanders, and service chiefs are directly responsible for security. So are the secretary of state and the heads of other departments and agencies. The intelligence staff officers in these departments lack the authority to issue directives and orders required for maintaining security. At the military staff level, security is an "operations" function, like a battle plan or organizational policy. In fact, operations plans normally have a security operations portion. Heads of intelligence organizations are responsible for the security of those organizations precisely because they do have full directive authority over them.

To repeat a practical example, Soviet penetrations of the U.S. embassy in Moscow were not the responsibility of any part of the Intelligence Community to stop. They were primarily the ambassador's responsibility and ultimately the secretary of state's. The Intelligence Community's responsibility was to discover the penetrations, to provide technical advice on how to eliminate the technical penetrations, such as listening devices, and to suggest changes in personnel policies the State Department might institute to prevent them.

Counterintelligence is also sometimes treated as a special function. It is different from intelligence in some regards, but it is also subsumed within the four primary functions. Counterintelligence is intelligence about an adversary's intelligence capabilities, targets, means, techniques, and so on. It is a subset of intelligence analysis and production, a special kind of intelligence product. Counterintelligence is used by commanders and agency heads to support their security practices, to make them more effective. And it is used by law-enforcement agencies to arrest enemy agents and to prosecute them. It is also used by commanders and policymakers to support "deception" operations and "covert actions." Counterintelligence has its special features, but so do many other types of intelligence analysis and production.

Counterintelligence depends on collection directed at the adversary's intelligence capabilities and activities. Collection managers must direct collection toward these targets. The results are used for counterintelligence analysis and production.

"Covert action" is not, properly speaking, intelligence. Traditionally, of course, the DCI has directed covert action, and only the CIA's Directorate of Operations has carried it out. Yet covert action actually constitutes "operations" in the same sense that the aims of foreign policy are implemented by diplomatic operations, and that military objectives are achieved by military operations. Covert action involves undertaking actions that cannot be directly attributed to the U. S. government. Covert action is meant to influence events without the objects of the influence being aware of the actual source of the influence.

Covert action is normally divided into two general types, paramilitary and all other covert operations. Covert action includes false or misleading information ("black propaganda"), using agents of influence, and many other political means; no force is involved. Paramilitary covert action involves training and using paramilitary units, and sometimes regular military units, for combat operations, most often insurgency operations.

Covert action can be conducted when the United States is at peace or when it is at war. In peacetime, paramilitary covert action has always been under the DCI's direction. In wartime, the picture has been muddy. The Defense Department's special operations forces are trained for paramilitary covert action. In the Korean War, the DCI and the U.S. Army were both involved. The same was true during the Vietnam War, the Persian Gulf War, and the war in Afghanistan. In principle, the DCI is supposed to transfer control of all paramilitary covert action to the unified commanders. In practice, control has remained a disputed matter, as was seen during the Persian Gulf War and probably during the war in Afghanistan, though reports are mixed.

Principles for Organizing to Perform Intelligence Functions

Collection management means different things in different parts of the Intelligence Community, but primary responsibility for it belongs to the intelligence analysis and production staff in all departments, agencies, and military units that use intelligence. That is where intelligence needs are determined, and therefore it makes sense to put collection management under the command responsibility of the users. No one else in the Intelligence Community is in a position to know better what intelligence users need.

Within specialized collection agencies and organizations, a somewhat different "technical collection management" function is required. There, managers must take the collection requirements coming from intelligence-using organizations and translate them so that the collectors know what is required. These managers must orchestrate the collectors to best exploit their capabilities. Technical collection management is a highly complicated and specialized activity, and it varies radically among organizations, depending on capabilities and the technologies involved.

In the military services, and sometimes in civilian departments and agencies, the intelligence staff "general collection" managers are painfully ignorant of the character of "technical collection" management within collection agencies, leading to confusion about who is really giving what collection direction. Clarification of the two types of collection management, therefore, is absolutely essential.

Collection

Three major and distinct intelligence-collection disciplines have emerged, according to the way information is gathered: human intelligence is gathered by agents working on site, signals intelligence is gathered by intercepting various types of communications, and imagery intelligence is gathered via photography or other image-capturing technologies, from land or sky. These collection disciplines long ago became so technically specialized that they had to be organized separately. No single collection organization can easily manage the training and operation of all three disciplines.

Human intelligence can be further divided into overt and clandestine collection. The skills and techniques for clandestine human intelligence, of course, are quite different from those required for overt human intelligence, and therefore designing organizations and specifying responsibilities for each is likewise different. So too is organizing the reporting and distribution of the intelligence collected by each means.

Signals intelligence began as communications intelligence—

for example, collecting messages sent by light, then later by radio and telegraph. As other electronic emanations, such as radar, emerged in World War II and later, collection of their signals was undertaken and exploited for intelligence. Electronic intelligence became the term for this part of signals intelligence. The spectrum of signals emanations now extends to include visible light, and a number of light signals have emerged more recently—laser emissions and electro-optics, for example—that can be collected for intelligence purposes. This technology—called measurement and signature intelligence—has raised questions about how such collection should be done, by whom, and how it should be managed.

Imagery intelligence began as photography, but in the past few decades, other imaging techniques have been added. Cartography and mapping have come to depend heavily on imaging from intelligence collection systems in space. Thus an argument can be made for lumping military and other mapping within this collection discipline. Technology has already pushed it quite far in that direction.

At the national level, although some signals intelligence, especially the electronic component, is highly fragmented organizationally, it has been largely concentrated under the roof of the National Security Agency (NSA). The trend since the early 1950s has been to increase the NSA's "operational control," as distinct from direct command and organizational ownership, over more and more signals intelligence collection and processing. In other words, each military service has large signals intelligence organizations that can, and do on occasions, "operate" independently without coordination and direction to prevent overlaps, gaps, and other inefficiencies. The NSA does not "command" them, but it can and frequently does direct their collection to ensure a proper division of labor and a more effective targeting of an adversary's communications. The turf boundary for operational control has always been a source of dispute between the NSA and the military services. If a military commander has a good technical understanding of signals intelligence, the dispute is less contentious, but many lack that understanding.

The signals intelligence model for distinguishing between command and operational control suggests an effective approach for cross-agency and cross-organization management of collection operations. It is the only way that orchestration of imagery intelligence collection can successfully be brought under a single national-level management structure. And it probably offers solutions to the traditional problems of coordinating the CIA's clandestine and overt human intelligence with Defense Department human intelligence.

At the national level, clandestine human intelligence has long been centralized in the Directorate of Operations in the Central Intelligence Agency. The Defense Department has had fairly large clandestine human intelligence capabilities, but its operations are at the coordinating discretion of the CIA Directorate of Operations. In principle, the Directorate of Operations has the same kind of operational control over military clandestine human intelligence as the NSA has over the military signals intelligence operations, but the organizational culture of human intelligence organizations has made it difficult to achieve full cooperation between the military and the CIA Directorate of Operations. The coordination between the NSA and military signals intelligence, however, has a lot to offer the national-level management of clandestine human intelligence as an example.

Overt human intelligence is carried out by a plethora of agencies and organizations in the Intelligence Community. Debriefings of defectors, travelers, and many other potential sources are conducted by CIA, DIA, and the military services. Collection of information from foreign news media, now done by the Federal Broadcast Information System, is in the Intelligence Community. The defense attaché system, consisting of military attachés residing in most U.S. embassies abroad, is another overt human intelligence collector, managed by the Defense Intelligence Agency. Prisoner of war interrogation units, which constitute yet another overt human intelligence collector, are found within the military services. Diplomatic reports, subscriptions to foreign publications and purchases of foreign books, information gleaned from meetings between U.S. and foreign officials, and several other kinds of overt human intelligence are outside the Intelligence Community. The degree of exploitation of these overt sources varies widely and is normally low. The whole of the overt human intelligence collection discipline lacks a coherent management structure.

The organizational principle advocated here is a system of "national managers" for each of the three collection disciplines. Until the Central Imagery Office was formed in 1992, creating the appearance with no substance, imagery intelligence lacked any semblance of a managing office. The Community Management Staff previously tried to manage imagery intelligence through one of its committees. As communications and imagery intelligence technology have changed, this committee structure has proved increasingly ineffective. Even the creation of the National Imagery and Mapping Agency in 1996 has not completely clarified the operation of imagery intelligence assets. Overt human intelligence has been almost as poorly managed, but because the public media, especially television, have become increasingly important, the deficiencies in overt human intelligence management are not felt so acutely. The news media compensate immensely.

Analysis and Production

Analysis and production are best performed close to the users of the intelligence product, normally by intelligence staff sections in the user organization. Exercising collection management, they can draw on the collection agencies to provide raw or processed intelligence for analysis, tailoring their products precisely for the users' needs. Alternatively, or additionally, processing (analysis and production) can be performed at centralized or separate analysis centers, apart from the intelligence staff sections of user military organizations and their equivalents in civilian agencies. Large central processing centers have greater capacities, but they tend to be less responsive to precise user needs. The two approaches can be mixed by augmenting the intelligence staff sections with separate supporting analysis and production units. For some types of analysis and production, large national centers may make sense, but the principle of analysis and production by intelligence staff sections is obviously preferable because it is better located to be attuned and responsive to users' needs.

Dissemination

Dissemination, of course, involves getting usable intelligence products to the appropriate users. It operates in response to the demands of collection managers and includes timely distribution from analysis and production staff sections to the appropriate users.

Dissemination is at root a communications issue. It includes the personal interaction between intelligence staffs and their users. And of course it includes the electronic or any other message system between collectors and producers, between producers and users, and sometimes directly between collectors and users when a single-source collection can be used without additional analysis or processing.

Command and management principles in military organizations sometimes interfere with rapid dissemination of intelligence. When intelligence derived at the national level through human intelligence, imagery intelligence, or signals intelligence collection is needed and usable at the tactical level in a unified command, it may be delayed in reaching a needy user by higher intelligence staff echelons as it passes downward. With modern communications and with the rapidity of operations at the tactical level, direct dissemination is often the best solution.

This aspect of dissemination is also related to the question of "dedicated" intelligence communications nets—that is, communications channels that can only be used for a single purpose, such as highly classified intelligence. Military communications experts tend to oppose dedicated nets because they limit overall flexibility in managing allocations of frequencies and systems, not to mention incurring additional costs. Considerable progress has been made in overcoming the conflicts between mutually exclusive principles for intelligence dissemination, but it remains a dynamic and challenging problem.

Counterintelligence

Although counterintelligence is not, properly speaking, an independent intelligence function, it has uniqueness that requires treatment as if it were. First, counterintelligence involves operations within the United States and within the ranks of the military services and civilian departments. Thus it straddles the foreign and domestic boundary. For collection targets it must look within the very ranks of the Intelligence Community. Second, it may lead to indictments and prosecution in courts of law. That involves it with the U.S. judicial system in a way that no other part of the Intelligence Community is fully involved. Third, its targets may include U.S. citizens who have no formal government affiliation. Fourth, while military commanders and civilian department heads use counterintelligence products to support their security operations and practices, they are also used for counterintelligence operations and clandestine human intelligence operations, by double agents, for example.

All of these characteristics of counterintelligence argue for a management organization with considerable autonomy. At the national level, an overall picture is essential if the counterintelligence effort is to guard against vulnerability to hostile intelligence. The first principle, therefore, is that a national-level manager, having operational control over counterintelligence within all departments and agencies, is essential.

A second principle concerns the connection between counterintelligence and law enforcement. Because espionage against the United States is a crime, some counterintelligence leads to law-enforcement operations. At the same time, catching spies and uncovering foreign technical collection capabilities within the United States (as well as abroad) are activities more complicated than catching domestic and foreign criminals. The motivations and resources backing criminals are different from those backing foreign intelligence services. Criminal investigation skills, therefore, often work poorly in counterintelligence operations. Counterintelligence and law enforcement are mixed organizationally in the Federal Bureau of Investigation, in the air force's Office of Special Investigations, and in the navy's Naval Investigative Service. Only in the army, where the Criminal Investigation Division and counterintelligence units are separate, is counterintelligence clearly differentiated from law enforcement in an organizational sense. The CIA's counterintelligence operations are entangled with its offensive human intelligence, another mix of operations that has drawbacks. Strong arguments for mixing offensive human intelligence and counterintelligence in a single organization in some cases can be made, but the arguments for mixing counterintelligence and law enforcement against ordinary criminals are not compelling. This set of issues needs effective examination.

A third principle for counterintelligence concerns its support by collectors. Counterintelligence organizations naturally want to run their own clandestine human intelligence—for example, double agents and penetrations. This is commonly understood. Counterintelligence exploitation of signals intelligence collection improved in the 1980s, but it can still make progress. Use of the entire array of imagery collection has not made the same progress. A multidisciplinary approach to collection for counterintelligence is an imperative principle in today's world of changing technology.

Covert Action

The overall utility of covert action has been a hotly disputed issue. Here we will set it aside and assume that the U.S. government will want to retain at least some capabilities for covert action.

Nonmilitary covert action probably has no other logical organizational disposition except within the clandestine human intelligence organization. Legally, however, the boundary between human intelligence and covert action has to be kept extremely clear. Influencing a foreign government without letting it know the source is a tricky business that requires clandestine access and means. That virtually marries most of it to clandestine human intelligence.

Paramilitary covert action is a special case, and one for which a considerable experience base exists. Two major problems have always beset it: who performs it and who controls it? The first question has an obvious answer. High-quality and extremely competent personnel are needed to perform it. Whether the Department of Defense Special Operations forces supply such personnel and equipment or whether the clandestine human intelligence service builds its own capabilities has long been the disputed question. It does not really matter which agency does it as long as the quality of the capabilities is high. In practice, however, the CIA has never given paramilitary capabilities a sufficiently high priority to ensure high quality. The increasingly technical aspects of covert action make this deficiency greater today than in the past. Paramilitary operations, of course, are primarily "military," not clandestine human intelligence. Thus the professional expertise for paramilitary operations in the CIA has not been great, especially in overall conception and command. Military affairs have grown far more complex, and keeping up with them has been difficult for the CIA.

The second question concerns control. Clearly in peacetime, the DCI, with his authority over clandestine human intelligence, is the proper person to direct and control such operations. In wartime, however, the need to integrate paramilitary operations within theater military operations puts the unified commander in the best position to control paramilitary covert action. He also needs to have a say in nonmilitary covert action during wartime. The official government policy has generally followed the principles suggested here, but the DCI and CIA have not always been willing to follow that policy. This issue needs to be settled, once and for all.

During the war in Afghanistan the Pentagon's Special Operations forces were used extensively. The full story of CIA-DOD relations in this case will not be available for some time, and even then it will remain distorted by both organizations. Initially the CIA was wholly without contacts and agents within Afghanistan who could meet Special Forces teams and introduce them to anti-Taliban groups. Reportedly, however, Russian intelligence services provided a number of sources to the CIA and the Defense Department. Judging from the few newspaper accounts of Special Forces teams' experiences, the CIA effectively used the Russian-provided sources for link-ups between Special Forces teams and Afghan groups, but many informal reports have drifted back from Afghanistan suggesting that many teams were met by no one. Remarkably, most of them struck out on their own, found anti-Taliban groups, and won their confidence for joining their efforts in the war.

A careful study of this experience in Afghanistan would be extremely useful in working out better arrangements between the CIA and unified commands. Whether or not it could be done without excessive organizational biases and distortions is an open question. A clue to the emerging biases against an objective review can be seen in the extensive media coverage the CIA allowed of its initial role in the campaign in Afghanistan compared to the scanty coverage the army has allowed of its Special Forces teams.[1]

In any event, covert action is not an intelligence operation. It is either a policy operation or a military operation. The Intelligence Community, of course, has to support it. Within the Intelligence Community, the distinction needs to be redefined, bringing the secretary of defense, the Joint Chiefs of Staff, the secretary of state, and the White House into the primary control role. Policy and military operations are their primary responsibility. When they turn it over to the Intelligence Community, ambiguities are inevitable, and responsibilities remain unclear.

Information Warfare

Information warfare, a recent coinage of considerable future importance, overlaps in some ways with covert action and deception operations. It therefore presents another organizational and policy problem yet to be sorted out. I lack sufficient practical experience in information warfare to generalize, but the issue needs to be on the Intelligence Community agenda. It clearly is an operations and policy function, as opposed to an intelligence function, but it requires enormous and complex intelligence support. As yet, the Intelligence Community simply does not have the technical skills for providing needed counterintelligence against information warfare.

Deception

Deception operations are important, especially in wartime military operations, but not necessarily confined to them. They depend heavily on intelligence capabilities of all types, and that sometimes leads to treating them primarily as intelligence operations, especially as clandestine human intelligence operations. Yet they can seldom be contained within the clandestine service because key officials outside of the Intelligence Community may have parts to play. They can be viewed to some degree as covert action, but are "deception operations" a form of "covert action"? The military services have not thought so. This is an ambiguous area yet to be fully sorted out.

Security

Although security is clearly a command and management responsibility, it requires strong intelligence support, and strong counterintelligence of the type most organizations do not really understand. Again, the example of technical penetrations of the U.S. embassy in Moscow shows the weakness of the security system. Technical counterintelligence capabilities in neither State nor the CIA were adequate to discover them, much less anticipate them. The NSA finally discovered the breach. Thereafter, some NSA and Defense Department officials insisted that they should dictate the security measures that the Secretary of State should take. The unfortunate quarrel arose primarily because there is no authoritative doctrine that defines who is responsible for security.

Development and production of communications security and computer security are the responsibility of the NSA, and thus under the aegis of the Intelligence Community. Both types of security represent huge challenges because of rapidly changing technology in the civilian community and abroad. Difficult legal issues confront both. Interagency turf disputes have been fueled by the questions of well-intentioned congressional committee chairmen, such as then-Congressman Jack Brooks's effort to assign a large part of responsibility for computer security to the Commerce Department.

I treat covert action, information warfare, deception, and security only cursorily here. Spelling out their place in this doctrine for intelligence operations is necessary, nonetheless, in order to show why I have largely excluded them. They transcend the Intelligence Community, although the Intelligence Community has an important role to play in their support and even execution in some instances.

Resource Management

Resource management concerns the acquisition and use of funding. It can be broken down into two separate but related functions. The first comprises programming and budgeting efforts, through which budgets are drafted for submission to the Congress. The second is budget execution—that is, spending the monies appropriated by the Congress in accordance with the laws authorizing the appropriations. In addition, the Intelligence Community is called upon to defend the current program budget before the Congress. All Intelligence Community agencies are deeply involved in resource management.

The DCI controls the first function, the programming, budgeting, and presentation to the Congress of the budget for the entire national-level Intelligence Community, which is known as the National Foreign Intelligence Program. In this process, he presides over a labyrinth of separate programs within the Department of Defense, the CIA, and the FBI, and several other departments and agencies. He is supposed to aggregate them in ways that eliminate overlap, improve efficiency, and ensure that all requirements for intelligence support throughout the government will be met. He is not responsible, however, for the huge tactical intelligence program budgets of the military services. They fall outside of the National Foreign Intelligence Program and are programmed within the Defense Department, first by each military service, then as "defense" or "joint" programs. We should take note of them here but need not sort them out. Complete compatibility between the tactical intelligence programs and the National Foreign Intelligence Program has never been accomplished effectively, though many attempts have been made and occasional progress has been achieved. The turf boundary between these tactical programs and the national programs is not confined to the Pentagon. It is also found in Congress, where the armed service committees oversee the tactical programs and the intelligence committees handle the DCI's programs.

The second function, budget execution, is left to the departments and agencies. The DCI has had his say in how budgets will be spent during the programming process before they go to Congress. Once the Congress makes its appropriations, each statutory Intelligence Community agency, or its parent agency, actually spends the budget under the rules of its parent department.

This system under which the DCI builds a consolidated interagency program and budget proposal has evolved over time, giving the DCI the authority to prevent duplications among departments in intelligence resource management. How effectively various DCIs have used this authority is an important issue for investigation, but in principle the DCI has the authority to allocate resources as they are needed by the various departments.

In the course of drafting reform legislation, Congress has sometimes considered providing the DCI full authority to execute the entire budget.[2] Although the goal of strengthening the DCI's powers over the Intelligence Community has some appeal, aside from the possibility that such an arrangement might be ruled illegal if reviewed by a court, it is bureaucratically impractical. It would create duplication of comptroller and auditing functions within several cabinet-level departments and the military departments within the Defense Department. The inevitable turf disputes and management complications would far outweigh any advantages to the DCI. The major task for contemporary Intelligence Community reform is to make more use of program authority that the DCI already has rather than expanding that authority to include budget execution. All the principles of resource management discussed henceforth concern programming, not budget execution.

The vertical rationalization of programming of resources— finding an efficient way to link national intelligence needs with tactical intelligence capabilities—is another matter. Quite naturally and logically, the Office of the Secretary of Defense dominates the programming of intelligence resources at the department-wide level, and the military services dominate the programming of resources at the tactical level. In the military services, it is often difficult to distinguish between resources strictly tied to intelligence activities and resources that also pertain to combat operations. A naval warship can collect intelligence; so can a fighter jet. In ground operations, rifle squads may be dispatched on reconnaissance missions. Entire companies, or even battalions, can be deployed as a reconnaissance force to locate enemy forces, effectively performing intelligence collection. Target acquisition systems for artillery and air-support operations—for example, unmanned aerial vehicles, ground-based sound-sensor devices, and electro-optic systems that see great distances—are not normally thought of as intelligence systems, but they can collect intelligence. Managing the programs within the Defense Department's tactical intelligence budget, therefore, involves dealing with ambiguous situations in which capabilities cannot be clearly defined as either intelligence or operational. The DCI's program management authority has never extended downward to account for capabilities in tactical intelligence.

I will touch on the need for improvements in integrating all these resources, but that is not my central concern. Trying to sort out all the complexities of an integration plan would require a separate analysis. One point, however, is axiomatic: until greater resource management rationality is achieved within the Intelligence Community in the National Foreign Intelligence Program, progress in integrating the tactical intelligence capabilities with national capabilities will be erratic and more by chance than design. Many observers and critics, especially members of the intelligence and armed service committees in Congress, have long noted the very large sums of money annually invested in tactical intelligence programs—Tactical Intelligence and Related Activities (TIARA), the Joint Military Intelligence Program (JMIP), and the Defense Airborne Reconnaissance Office (DARO)—and the lack of coordination with spending on national intelligence programs.

The key issue for resource management is understanding how money spent and personnel deployed relate to collection of usable intelligence. Since the early 1960s, when Secretary of Defense Robert McNamara introduced the planning, programming, and budgeting system in the Pentagon, there has been a general recognition of the need to relate resources more effectively to missions—that is, the input-output relation. By law, budgets approved by Congress are organized on the "line item" principle. A line-item budget lists tanks, aircraft, belt buckles, boots, hats, radios, and thousands of such things without grouping them in support of any purpose more specific than "defense of the United States." Program budgets under the system that McNamara introduced make a more rational connection between military budgets and output capabilities by aligning the line-item budget so that each item appears under a specific mission area: conventional forces, strategic forces, homeland defense, reserves, and so on.

Because the DCI has never made the effort to impose a similar system on resource management in the Intelligence Community, its consolidated Intelligence Community budget does not effectively relate inputs to outputs. Although budget meetings of the Intelligence Community Executive Committee generally include some discussion of principles similar to those of the Pentagon system, no management and programming system has been established to evaluate input-output relations.

How can the planning, programming, and budgeting system be adapted to the Intelligence Community? The obvious way to begin is to look at the main functions specified for the intelligence cycle: a) collection management, b) collection, c) analysis and production, and d) dissemination. Two of these, a) and d), are command and control functions. The final output function is c). The d) function provides the most important measure of the Intelligence Community's output, but with a qualification. The output disseminated cannot really be counted unless it is used by policymakers and diplomats and by military commanders and their operational staffs. The intelligence process often produces answers to questions that no one is asking or in forms that cannot be used—data that cannot rightly be counted as outputs.

Thus the major resource management question for the DCI is whether the Intelligence Community's output is used and actually meets the needs for policy and operations. And the Intelligence Community organizations in the best position to answer that question are the staff intelligence sections in all military units and in the civilian departments and agencies whose operations rely on intelligence.

It also follows that if intelligence production is to be usable, relevant, and sensitive to the needs of users, then it must be done close to, and in constant working contact with, the users and their staffs. The military paradigm here is instructive. The concept of a staff intelligence officer assumes that collection management, analysis, production, and dissemination are accomplished by this officer, or by the equivalent staff sections in air force and navy units. The same idea is behind the existence of an intelligence section in the State Department. Similarly, most civilian departments and agencies have staff intelligence sections.

The White House as a user of intelligence is a special case. According to a popular misconception, the National Intelligence Council, supported by the CIA's Directorate of Intelligence as a kind of national-level chief intelligence office, filter intelligence and submit the final, fully analyzed product to the president, who reads it and bases policy decisions on this output. In fact, to the extent there is an intelligence staff section within the White House, it is the National Security Council staff. Staff members integrate most of the available intelligence—both raw information from various sources and also some analyzed intelligence—with White House policymaking. Finished intelligence products periodically do reach the president, but these represent exceptional and usually marginal influences on his decisions. The daily deluge of raw and finished intelligence that the National Security Council staff sees represents the major influence that the Intelligence Community has on the president's view of the outside world. And even this competes with the mass media, individuals inside and outside gov-

ernment whom the president consults, and foreign leaders and officials who give him their views and other information about foreign, economic, and military affairs.

I have yet to discuss the collection function with regard to
input and output measurement. Collection is best considered by
discipline: human intelligence, imagery intelligence, and signals
intelligence. From the standpoint of resource management, each
can be treated as an autonomous input-output entity. And counterintelligence collection is sufficiently distinct in many of its aspects to demand separate treatment as well. Applying the principles of the planning, programming, and budgeting system, each of
the disciplines can be seen as a program-mission area, where input
can be related to output.

The managers of human intelligence are in the best position
to make judgments about what levels and mixtures of resource inputs produce more or less output of human intelligence. The same
is true for imagery intelligence and signals intelligence. The measurement of output, however, cannot be the prerogative of these
managers but should fall to the policy and military users. In other
words, resource managers face a large community of customers,
nationally throughout the civilian and military departments, and
also through the lower levels in the military system of unified commands.

Budgets for the Intelligence Community, like military budgets, are broken down into three categories: operation and maintenance; procurement; and research, development, testing, and
evaluation (R&D). Congress insists on these categories, and the Intelligence Community must operate within their limits.

Management of the three budgetary categories falls to separate and sometimes autonomous subunits. Managing R&D requires different skills and organization than does managing operation and maintenance. Procurement also requires specialized

organization, skills, and management. Intelligence outputs, however, depend on inputs to all three. Applying the planning, programming, and budgeting system principle, then, at various levels of organization, management must work across all three categories, relating inputs for all three budgets to intelligence output. Inevitably, organizations that carry out only R&D and procurement develop their own subunit interests that conflict with the interests of intelligence operations organizations and their operations and maintenance budgets.

Here we run into a major organizational and management tension between R&D and procurement organizations and those that primarily undertake operations, spending operation and maintenance funds. The input-output relation, which links them, conflicts with the internal organizational dynamics of each. Over time, parochial bureaucratic self-interest in each type of organization will begin to undermine both overall efficiency and the ability of individual organizations to make the best use of limited aggregate budgets.

It should be noted that not all Intelligence Community organizations and their subunits can be neatly labeled as purely R&D, procurement, or operation and maintenance. All have some operation budget, some have a procurement budget separate from operation and maintenance, and some have all three. Still, R&D and procurement tend to be funded disproportionately in specialized organizations that are not primarily collection organizations. A serious organizational pathology is the tendency of R&D and procurement organizations to usurp collection operations such that managers in collection organizations have inadequate control of these operations. Likewise, organizations devoted to collection may respond by trying to conduct their own R&D and procurement.

A major bureaucratic cause of these organizational pathologies is the process by which programs and budgets are developed and approved. A fixed total figure for the budget is established at the beginning of the process. Each organization competes for its share in a zero-sum game. The best mix, of course, depends on the kind and quantity of intelligence output needed to satisfy users. But competition for budget dollars can shove that consideration aside. Often an organization judges its own success by the size of its budget. Those intelligence organizations most closely tied to intelligence users—typically staff intelligence producers and the suppliers of collected raw intelligence—also sometimes face pressures to allocate resources in ways that are at odds with the zero-sum game of the programming process. R&D and procurement organizations, which have the least direct responsibility to actual users of finished intelligence are much freer to pursue their own budgetary goals.

The National Reconnaissance Office provides the most egregious example of this pathology. Wholly a research and development and procurement agency, it has traditionally striven to maximize its budget, often spending several billions of dollars that were impossible to justify in a clear input-output analysis. As a result, it has repeatedly squeezed out small programs that possibly offered highly preferable outputs of intelligence.

There are no management panaceas for these tensions and dysfunctional behavior patterns. In general, however, single management across the three budget categories is preferable at levels where the input-output relationship across the organization—and thus appropriate priorities among the budget categories—are easily assessed. The greater the separation of organizational structures for R&D, procurement, and operation and maintenance, the more difficult and less effective will be the management of the input-output relationship.

As a general rule, managers at intermediate and lower levels have better prospects of improving input-output relations if they have control of all three budgeting categories within a collection discipline. At the same time, managers at these levels often refuse to take significant R&D risks and adopt a longer view because they feel the daily pressures for delivering intelligence to analysis organizations and, in some cases, directly to users. Their view of the long-run future, when R&D investment may pay off, is like that of Lord Keynes: "In the long run we are all dead."

Subordination and Autonomy for Intelligence Agencies

Another doctrinal issue of fundamental importance for intelligence reform concerns subordination and autonomy. Do the nature and importance of a particular intelligence activity demand special organizational autonomy for the agency involved? Or is the agency likely to be more effective when its performers are organically subordinated to user organizations? Arguments for organizational autonomy are made only by the CIA, although occasionally NSA officials have advanced them.

Organization theorists would probably conclude that wholly independent intelligence agencies are more vulnerable to parochial self-interest that undermines the primary goal of supplying timely and usable intelligence. Agencies that are directly subordinate to users—and that rely on those users for budget support—are more likely to be responsive to those users.

To look at the point another way, suppose that single-family households were forced to sign up to a household maintenance organization for all their needs—electrical repair, plumbing, roofing repair, painting, lawn maintenance, and so on. Rather than a private-sector relationship between service buyer and service seller, suppose that this maintenance organization went to the Congress for its budget, and the heads of the households had to pay a federal tax to support federal funding of the maintenance organization. The maintenance organization would be pushing for higher budgets, and the heads of households—a diffuse group not easily organized into a lobby—would be opposed to higher budgets. The intense and concentrated lobbying by the maintenance organization would probably defeat the heads of household lobby. The real-world example of this arrangement is Medicare. The Medicare budget began at a modest level but soon developed a dynamic beyond easy political or organizational control. This is precisely the dynamic to be expected from highly autonomous intelligence organizations.

This kind of dysfunctional organizational behavior is so predictable that one wonders why autonomy has been granted to intelligence organizations in a few cases. Yet there are reasons for it.

The oldest argument is based on the assumption that subordinated intelligence organizations are subjected to biases in analysis and reporting dictated by their users. The failure of military intelligence organizations to anticipate the attack on Pearl Harbor, it will be remembered, was cited after World War II as an example in support of making the CIA wholly autonomous. There is something to this argument. Users, and not just military users, do not like to receive intelligence analysis that contradicts their preconceptions and favorite policies and plans. In wartime operations they tend to be more attentive to bad news because they may face defeat in combat if they reject candid and valid intelligence. Still, commanders and political leaders have often ignored accurate but unhappy intelligence findings. But intelligence organizational autonomy is not a sure remedy to this problem. Autonomous intelligence organizations themselves take on biases in their analysis, creating another source of distortion.

No organizational solution to these biases—user biases or intelligence analyst biases—offers a panacea. They are usually corrected either when an adversary's behavior eventually exposes an intelligence bias or when an analyst or intelligence official defies institutional biases to make a case based entirely on a revealing analytical interpretation. Only the first corrective, the adversary's behavior, is completely reliable—and sometimes extremely costly. The second corrective, honest and insightful intelligence analysis, is problematic. More objective officials will arise if the intelligence culture encourages them, fewer if it discourages them. But in some cases intelligence officers assert the need for autonomy to defend their integrity while at the same time providing poor analysis, characterized by lack of insight and simple misunderstanding of the evidence. Autonomous intelligence agencies will always be vulnerable to such people, despite their honesty and good intentions.

Another corrective to bias often proposed is competitive analysis: the A Team versus the B Team, each investigating the same matter independently. Proponents of competitive analysis sometimes cite the example of the academic community, where competing scholarship supposedly thrives in an honest marketplace for the truth. Leaving aside for a moment the putative success of this model in academia, this corrective has yet to demonstrate its efficacy for intelligence, though it has long been institutionalized in various agencies. Competitive analysis has seldom produced better analysis, but it has frequently inspired intense parochialism. As a rule, it creates more heat than light.

None of the traditional arguments—from the Pearl Harbor example to the academic model of competitive scholarship—makes a compelling case for autonomous intelligence organizations. The academic argument, for one, is based on a misunderstanding of the nature of scholarship. Truth in scholarship depends on a combination of individual integrity and genuine insight. It thrives not because universities offer an exceptionally effective marketplace for ideas. Rather it thrives because senior scholars occasionally have the integrity and self-confidence to encourage younger scholars to do innovative and creative work, sponsoring them rather than competing with them. A similar dynamic can sometimes be seen in intelligence analysis units. A senior intelligence officer will recognize a subordinate's insight as valid although it conflicts with the senior's favorite answer to a question. Accordingly, the senior official sponsors and adopts a new, unconventional, wisdom.

The intelligence user can also encourage this kind of behavior among his intelligence staff officers. He takes a strong interest in the analysis but a dispassionate one, letting his intelligence analysts know that he is more interested in the unvarnished truth than in a preferred truth. He is tolerant of mistaken analysis if it is not defended in a parochial fashion, perhaps even rewarding the mistaken analysts who readily surrender their hypotheses when the evidence undercuts them.

No organizational technique or structure will ensure this desirable kind of behavior among either intelligence officials or users. But certain kinds of organizational management approaches can encourage it.

Are there no good arguments for autonomy of intelligence organizations? Yes, there are. The best ones are based on complex technology and specialization of skills. A very old example is cryptanalysis, the process of breaking communications codes. The number of people who can master cryptanalysis at a level that is productive for signals intelligence has always been small. H. O. Yardley's infamous American Black Chamber, set up by the army during World War I, was the first modern and effective cryptanalytic endeavor. It was a highly centralized and autonomous organization, but its products were distributed not just to military users. Diplomats became voracious consumers as well, and the 1921 Washington Naval Conference's arms control treaty was critically shaped by Yardley's products. In World War II signals intelligence based on cryptanalysis was highly centralized and remarkably effective. That effort, jointly shared with the British at Bletchley Park, England, although somewhat autonomous, was under the command of the Supreme Allied Commander, General Eisenhower. He and his staff were in a position to limit the dissemination and use of broken codes. Lower-level commanders enjoyed the advantages of the work at Bletchley Park; it would have been impossible for each to have had an organic cryptoanalytic unit capable of the same quality of output. Centralization and greater autonomy for that cryptanalytic organization therefore made sense. Still, signals intelligence was also produced at the tactical level by units at lower command levels. Commanders could direct these units' efforts wholly toward their own intelligence requirements.

Another example of organizational differentiation and autonomy at the *apparent* expense of the supported command is found in artillery. Any military commander prefers to rely on forces under his own command rather than on outside support because he expects his own units to be more responsive. Technology changes, however, frequently favor specialization and support relationships. As the range and accuracy of artillery increased, centralization of command at higher levels was essential for exploiting those technological changes. Brigades and regiments gave up command of artillery, and a new doctrine evolved to let the commanders at division and corps levels allocate direct or general artillery support as needed for their operations. Lower-level commanders did not like the change, but practice has proved its wisdom.

The changing methods of tactical air support and strategic aviation reflect an analogous organizational response to new technology. Extensive doctrine for providing close air support has been developed and used to reasonably good effect; ground force commands do not have—or need—their own tactical air units. Logistical, medical, and engineering capabilities, among others, have been similarly centralized.

These nonintelligence examples can be instructive for intelligence organization. A degree of autonomy and specialization in organization offers tremendous advantages if it is accompanied by changes in doctrine for operations to ensure support for the command levels that surrender direct control. Intelligence has been reshaped by vast technological changes and agencies have been allowed a great deal of organizational specialization and autonomy. Intelligence doctrine, however, has not kept pace to ensure effective support. As a result, some of the organizational specialization and autonomy has been counterproductive. I have made the distinction between command and "operational control" with regard to intelligence collection assets, and we have seen examples of the successful application of that distinction in signals intelligence. When specialized collection organizations provide intelligence support to users, the principle of general and direct support could be borrowed from the artillery. National-level collection and production of signals intelligence has been provided on a de facto principle of general and direct support, based on priorities determined by the chairman of the Joint Chiefs of Staff. The Joint Chiefs decide which unified commander is to receive priority in a crisis or for some other special and temporary operation, and then the National Security Agency allocates assets on that basis. This is precisely how division and corps commanders mass their artillery for greatest effect.

It must be emphasized that autonomy must be limited for effective intelligence. Somewhere up the chain of command, the most effective intelligence operations are subordinated to user commanders or policymakers. That principle has been jeopardized, however, by the creation of an intelligence agency at the national level, subordinated only to the White House: the CIA. The

White House is unlike any cabinet or other independent department. It has never been considered a managing agency but rather a policymaking agency that implements policy and operations through cabinet departments and agencies. There may be compelling reasons to maintain the CIA as one of those implementing agencies, largely independent of all the agencies it is supposed to support, but the problems with the CIA over the past quarter of a century call into question whether those reasons really are so compelling. The reluctance of DCIs to force CIA human intelligence to be fully responsive to military requirements, especially during crises, needs to be investigated. It may be that they simply do not have the administrative capacity to compel it. It may be that the DCI's role as CIA director encourages him to take a parochial bureaucratic view.

Perhaps if the DCI gave up his double-hatting as head of the CIA and, in conjunction with the Community Management Staff and some sort of analytic body such as the National Intelligence Council, he could stand above the CIA, allowing him to hold it to account rather than to serve as its parochial spokesman. The DCI has considerable power that does not derive from wearing his CIA hat. He has three kinds of power that have not been fully exploited. First, as the national program manager for the entire Intelligence Community, he can influence resource allocations. Second, his authority to "task" all intelligence agencies to collect according to the priorities of the national intelligence requirements lists, which his Community Management Staff compiles, is no small power. Third, his authority to coordinate and endorse the production of so-called national intelligence products, such as National Intelligence Estimates and many less formal analyses, which he does through the National Intelligence Council, affords great potential for setting the entire Intelligence Community's production agenda.

Organizational reform solutions will not be found in generalities like "competitive analysis" or references to Pearl Harbor. They can be discovered only through investigations based on in-depth knowledge of technology and operational and policymaking processes. And they require more than a little trial and error. At the same time, both investigatory analysis and practice can be guided by some underlying concepts.

The doctrinal principles sketched out here are meant to provide such concepts. They are not immutable or sacred. Rather, they have been arrived at inductively, based on what seem to be enduring effective paradigms of organization and functions in intelligence activities over the twentieth century. Just as "fire and maneuver" has remained an enduring principle in military operations even as military technologies have changed radically, the intelligence cycle and collection disciplines have remained continuously valid.

A final point about subordination and control concerns the military character of intelligence operations. Military examples have predominated in this effort to set forth a general doctrine, or set of organizational principles. That is not accidental. Intelligence has always been closely, even primarily, related to military operations. Diplomacy also has long been associated with intelligence, but at root, intelligence operations have a closer affinity to the military ethos than to any other. Both are competitive, and, despite the limited success of attempts to regulate wars through international norms and laws, both generally operate outside traditional legal systems. Spies risk and lose their lives, just as soldiers do. The same is not so true for diplomats, although they take exceptional risks at times.

Civilian control of the military is a fundamental axiom of the American political system. Given the military-like character of intelligence operations, to what extent is that axiom valid for intelligence organizations outside the military services and the Department of Defense? To the greatest degree. The issue, however, is not whether spies report to uniformed military officers. It is really a matter of political accountability. Civilian bureaucrats and lower-level political appointees in the Defense Department are not the direct source of civilian control over the U.S. military. The ultimate control, though, is political: responsibility before the electorate. Congressional control over the budget is the core element of that support, and the power of the voters over the president and the Congress stands behind the powers of the purse. Presidentially appointed officials in the Defense Department exercise the president's political control over the military just as the Congress controls it through the budget. The president does not personally know many of the scores of his appointees in the Defense Department, and his capacity to know whether they actually are fostering his policies is even more limited, probably reaching down no farther than a handful of people below the Secretary of Defense. Thousands of civil servants in the upper reaches of the Pentagon are not bound by the ethos of a military officer's responsibility to the Constitution and political authority, an ethos cultivated from the very beginning of an officer's training.

I make these observations to anticipate the common argument that the CIA, as a "civilian" agency, effects civilian control over intelligence operations that would not exist if they were carried out by military personnel. This is clearly a false argument. There is indeed a problem of political control over intelligence operations, but it has nothing to do with clothing, military uniforms versus gray flannel suits. In fact, the principles of military officer training—duty, honor, country, and dedication to the principles of political authority based on the Constitution—would be at least a psychological control factor for an independent intelligence agency like the CIA. Its civilian personnel, however, are not subjected to enough of this kind of socialization and character training before being posted as intelligence operators and analysts; the retraining they receive as their careers progress appears to be even more sporadic.

Political control over the CIA is therefore a serious issue, a growing problem, but not because military officers are in charge. The DCI and sometimes his deputy are the only political appointees acting for the president in controlling the CIA. This issue, therefore, deserves scrutiny, but not of the sort that is usually demanded.

Another variant of the issue of civilian or military control concerns the presumed propensity of military intelligence officers to shortchange nonmilitary intelligence needs. The grounds for this worry are difficult to find. The National Security Agency (NSA), according to DCI William Casey, supplied about 80 percent of all national intelligence needs in the mid-1980s. That percentage has been fairly consistent and probably still applies today. More than twenty civil agencies depend heavily on its collection. Few if any of them complained about being shortchanged. Yet the NSA is commanded by a military officer, and more than two-thirds of the personnel working under NSA command or operational control are military. The CIA and the FBI are wholly civilian agencies but have much poorer records of satisfying civil agency users. And the Defense Intelligence Agency (DIA), a military organization, has no record of turning down civil agencies' requests for intelligence. In fact, DIA has vigorously pursued the distribution of its products to the White House and to any other nonmilitary user who will accept them. All in all, the fear that military intelligence agencies will not respond to the DCI's instructions that products be distributed to nonmilitary agencies seems groundless.

Intelligence Management Training

Even if a common doctrine for the Intelligence Community were promulgated as official policy, the community's behavior would probably change very little without comprehensive and recurrent management training. Few senior intelligence officers—military and civilian—today know enough about the entire Intelligence Community structure and operations to apply a common doctrine. Senior management training is desperately needed. This applies not just to resource management but also to collection management. The intelligence officer on a navy ship can request and probably receive vast amounts of intelligence from nonnaval intelligence collectors. An army staff intelligence officer could do the same. Yet nowhere in the intelligence schooling system are officers headed for these assignments taught, or allowed to know, how to request national and other outside collection support. The same is probably true of the collection management and analysis elements in civilian cabinet departments.

All components of the Intelligence Community have training programs and education systems. Intelligence schools are especially numerous in the military, and the Defense Intelligence Agency runs the Joint Military Intelligence College and the Joint Military Intelligence Training Center. Would-be agents take a wide array of courses in photo interpretation, signals intelligence, clandestine trade craft, and special kinds of analysis. The Department of Defense's Defense Language Institute contributes to the Intelligence Community through its many language training programs, and several other foreign-language schools are available to the Intelligence Community. In short, the Intelligence Community's system of education and training is extensive, diverse, specialized— and fragmented.

Much of this school structure could be improved—made more relevant, raised in quality, made more comprehensive. That is a perpetual challenge faced by the intelligence schooling system, rather than an issue for systemic reform. Performance of various components of the schooling system ranges from outstanding to fairly poor. The technical collection disciplines are always struggling to keep up with changing technology. Human intelligence training, especially in counterintelligence, is uneven. Training in collection management is poor except in some narrow technical spheres within the technical disciplines. Training in analysis and production is also mixed in quality, with inadequate attention given to teaching analysts how to gain access to all sources of intelligence. Most good analysts learn them on their own.

Major Deficiencies

The Intelligence Community school system lacks three key elements. First, nowhere is a common doctrinal understanding of intelligence functions and processes documented and taught to all Intelligence Community management and executive leadership personnel. Second, the teaching of community-wide resource management has been generally neglected. Third, there is no educational emphasis on senior executive leadership and staff training.

Earlier I elaborated a number of doctrinal concepts that provide a common language for understanding and organizing intelligence operations. Today, if a dozen or so senior officials from all Intelligence Community components were pulled together and quizzed on these principles, they would not give compatible answers—or sensible ones, in most cases. For example, collection management has many different meanings in practice, and senior officials from different backgrounds bring quite different understandings to this term. The same is true for most of the intelligence functions and for more specific language within collection disciplines. The result is miscommunication within the Intelligence Community. Intelligence officers rise to high posts with extremely parochial and limited comprehension of intelligence functions and processes. Yet they most often assume that they know the essence of the Intelligence Community and intelligence as a profession. In fact, almost no one has a comprehensive view of the Intelligence Community and its operations.

This state of affairs in senior management can be explained in part by the requirements for security compartmentalization and limited, need-to-know, access to many Intelligence Community programs. An official who grows up in the human intelligence discipline inevitably has a limited comprehension of signals intelligence and imagery intelligence. Those with experience in tactical military intelligence understand aspects of intelligence operations that remain a mystery to more than 90 percent of senior management personnel in the Intelligence Community. Likewise, many tactical military intelligence specialists know very little about national collection systems or about providing intelligence support to civilian departments.

Even within the military, there is no standard doctrinal approach for joint intelligence staffs. Collection management is performed differently in each. Staff subfunctions also differ between the joint intelligence staffs of the unified commands. By and large, each is an ad hoc arrangement. Little wonder that most of these staffs have difficulty drawing on all intelligence collection capabilities available.

The lack of common doctrinal understanding in the CIA has kept it in counterproductive struggles with virtually all other components of the Intelligence Community. At times, the agency has tried to take over virtually all Intelligence Community collection management—in fact, the old Intelligence Community Staff had that function officially. The Defense Intelligence Agency has had similar turf fights with the National Security Agency with regard to the joint intelligence staffs. Counterintelligence, clandestine human intelligence, and imagery intelligence have never easily adjusted to support military operations, precisely because what such support entails is not specifically spelled out and taught as standard procedure.

Although this parochialism and limited knowledge may be understandable, they are wholly unacceptable among the senior ranks of Intelligence Community leadership. Until these failings are overcome, they will continue to obstruct meaningful dialogue and cooperation within the top circles of the Intelligence Community. A common doctrinal language has to be mastered by all senior Intelligence Community officials, and it has to be updated continually in light of new technology and new operational concepts based on practical experience.

No less important than a common doctrinal language for the Intelligence Community is a common understanding of community resource management. Earlier I raised a number of key issues for resource management. Such issues need to be made the focus for training and education of both lower- and higher-resource management officials. Here again, the absence of a common language and set of understandings is a serious deficiency which Intelligence Community education reform must address. This means more than technical training in program management. Training should also include critical analysis of the levels of efficiencies in the present resource management techniques and methods.

Finally, the Intelligence Community needs senior management and leadership training for its civilian personnel just as much as most large organizations do, if not more. The senior military personnel in the Intelligence Community have somewhat more leadership training in early- and mid-career schooling, but this training is not particularized to the Intelligence Community.

More serious than lack of leadership training in the Intelligence Community is lack of training in high-level staff work. It may be that most military officers in the Intelligence Community do not really comprehend either the terminology of staff work or the divisions of labor that that terminology implies. But the Intelligence Community's civilian personnel are far less schooled in all of these matters. Most have been promoted to senior positions because of their acute bureaucratic skills, not their management and staff work skills. Simple organizational processes and principles, common to virtually all organizations, are unevenly understood among the senior executive service civilians in the Intelligence Community. Many civilians are sent to the military services' war colleges, but these schools are totally inadequate for the senior leadership needs of the Intelligence Community.

Recommendations

- The DCI should create an Intelligence Community senior management education system.
- This system should have as its core curriculum three areas: a) Intelligence Community doctrine; b) resource management; and c) leadership and staff work.
- A simplified version of this curriculum should be required for entry-level and midcareer schooling of intelligence officers, as well as for senior-level officers.
- This schooling system should use senior line and staff personnel as instructors.

Getting a competent faculty for intelligence schools will not be easy. Incumbents in most senior positions in the Intelligence Community should be encouraged to devote significant time to teaching courses in the Intelligence Community school system. This management and leadership schooling is not a substitute for any of the current Intelligence Community education or training. Rather, it must be added to current education, and it must be focused and select at the top levels.

The many concepts, principles, and arguments presented here cannot be applied in a mechanistic fashion that will easily turn up all of the structural and procedural problems in the Intelligence

Community and provide solutions to them. They can, however, provide a common language for seeking out and distinguishing between remediable problems and unresolvable tensions and frictions. And they should, if pragmatically applied, lead to a set of recommended reforms that are consistent and mutually reinforcing, rather than ad hoc, incomplete, and perhaps even contradictory.

3

Making Dollars Yield Useful Intelligence

The inchoate outlines already exist for a management structure through which the director of central intelligence can direct and manage the Intelligence Community. Some of the parts of this outline appeared early, while others emerged over time. In the late 1960s through the early 1970s, substantial organizational attention was devoted to soliciting the intelligence collection requirements of all departments in the government, prioritizing them, and tasking various agencies to collect intelligence to meet them; other areas of progress included resource and program management and Intelligence Community–wide policies. From 1947 there was always a board (known today as the National Foreign Intelligence Board) that approved "national" intelligence analytical products. To support this board and the DCI's use of it, the National Intelligence Council was created in the mid-1970s, a sort of analytic staff composed of "national intelligence officers" who chair interagency groups drafting "national" intelligence products. At about the same time, a council (now the Intelligence Community Executive Committee) was created to deal with resource issues. A staff (today's Community Management Staff) was also formed to assist the DCI's use of the committee, as well as to handle a wide range of administrative and policy issues applicable to the Intelligence Committee as a whole.

In principle all of these developments were positive for building a coherent Intelligence Community, but in practice they have never reached full fruition. Some of the reasons for that outcome will become apparent as I assess the current arrangements against the backdrop of the concepts and doctrine in the previous chapter, but the main obstruction has been the double-hatting of the DCI as the director of CIA. The dual responsibilities make it difficult for the incumbent to rise above the organizational interests of the CIA and to act effectively as the director of the entire Intelligence Community. The dual responsibilities encourage the impression both at CIA and in all other intelligence agencies that the CIA, not the DCI and his community organizations, is in charge of the entire Intelligence Community.

The most persistent argument advanced against splitting the two positions is that without the CIA hat, the DCI would have no bureaucratic base with resources directly at his disposal. As I assess the current structural arrangements in this chapter against the principles and concepts of the previous chapter, it should become apparent that this old argument is not persuasive. If the recommended changes are implemented, the DCI will have an abundant bureaucratic base and much improved capabilities for managing the entire Intelligence Community.

DCI Management Functions

The DCI has three distinct roles. Each must be fully clarified and understood before any effective reform of the DCI's role is possible.

Collection Management, Intelligence Analysis and Production, and Dissemination

As the top intelligence official in the government, the DCI is responsible for producing intelligence analysis and judgments for the president and cabinet-level officials, and at the same time, he has to deal with the conflicting intelligence analyses produced by various components of the Intelligence Community. In other words, he should coordinate and support intelligence production not just in the CIA and the National Intelligence Council (NIC) but throughout the entire Intelligence Community. And he must have something like the National Foreign Intelligence Board with its senior membership to review and approve or take exception to national intelligence estimates and other national intelligence analysis.

Intelligence can be produced only if information is first collected, and thus the DCI has to provide collection management for the Intelligence Community. Part of that duty concerns non-time-sensitive collection for the entire federal government. That requires a system for collating the intelligence requirements of all departments and agencies, prioritizing them, and assigning them as collection tasks to appropriate agencies—that is, to those engaged in the three collection disciplines: signals intelligence, imagery intelligence, and human intelligence.

But collection management and intelligence production take place in scores of places, not just at the national level. Each cabinet department has its own special intelligence needs. Within the Defense Department, the needs for specialized intelligence products are different for the Joint Chiefs of Staff, the unified and specified commands, and in the military services. Below the level of the secretary of defense, virtually every military organization possesses its own intelligence staff section to respond to its particular intelligence analysis needs.

The DCI cannot possibly direct all of this collection and intelligence production. And he should not. The policy officials and military commanders (who know best what intelligence they need) must direct it themselves, assisted by their intelligence staffs. The DCI, however, can and should perform a general management role over most of this dispersed community of intelligence analysts. He can look for duplication, for analysis that is done without adequate access to collected intelligence, for lacunae in areas of analysis. And he can resolve disputes that affect national-level intelligence production. Finally, he can ensure that the president, through the National Security Council staff, is provided with whatever intelligence analysis he requests, as well as any unrequested intelligence analysis he believes the president needs.

As made clear by the principles established in chapter 2, the DCI's obvious staff support for collection management and intelligence production is the National Intelligence Council (NIC). As the Intelligence Community now operates, however, collection management is performed by the Community Management Staff. But because the CIA often assumes that it is the DCI's primary intelligence production staff, that agency has tried to run collection management for the DCI. Not surprisingly, many turf quarrels have arisen as a result. The National Intelligence Council as it is now constituted is both too small and ill-equipped to perform this expanded task. Rather than a small staff of national intelligence officers, it would have to become a significantly larger organization with a broader range of competencies. It would also have to separate itself from residency at the CIA, stand apart from all the main intelligence agencies, and strive to become a helper and sponsor for intelligence analysis groups throughout the entire government.

Resource Management and Policy for the Intelligence Community

Resource management and policy includes three distinct management responsibilities for the DCI. First, as the director of the Intelligence Community, he must evaluate the community's outputs. That requires a system for evaluating collection organizations' responsiveness to national intelligence requirements. Second, he must manage resources—that is, supervise Intelligence Community programming and budgeting and decide the allocation of funds and personnel. Third, he must make policies applicable to the entire Intelligence Community.

The obvious vehicle for assisting the DCI in executing these responsibilities is the Community Management Staff (CMS). The CMS cannot support the DCI adequately if it is merely a parliament for Intelligence Community components, as it traditionally was in its earlier incarnation as the Intelligence Community Staff. Personnel sent by various organizations to serve on this staff were expected to prevent it from effectively involving the DCI in their business.

The CMS must be truly a staff, neither a directive body nor a bureaucratic parliament designed to obstruct. It must identify and analyze problems, develop solutions, and present its findings to the DCI. Decisions rest with the DCI, and take on the power of his directive authority.

The DCI's responsibilities to evaluate outputs, manage inputs, and make policy must be understood in the interdepartmental and organizational context of the Intelligence Community. The DCI does not have authority to override or interfere with management and command within the Defense Department, the military services, or other agencies with complex and deeply embedded intelligence structures. The nature of the Intelligence Community makes tensions unavoidable between the DCI on one hand, with his responsibilities for management and particularly for direction, and the managers and commanders of various departments on the other. Such tensions, however, can be kept limited if the DCI is not too literal about "direction" below the upper levels of the Intelligence Community. In some cases, of course, his policies—for example, security classifications and handling of intelligence products—must be accepted as unambiguous directives. But in other, less clear-cut issues, if the DCI brings the heads of the Intelligence Community components along with him, they will help mitigate bureaucratic conflicts as well as make the DCI's policies and management aims penetrate more deeply inside all independent departments and agencies of the community. In fact, an effective style of leadership can help the DCI preside over an Intelligence Community that really *is* a community.

Intelligence Community Coherence

If the Intelligence Community is to reflect a genuine sense of "community," the DCI has to be responsible for asserting, fostering, and exploiting that sense. A council or committee of the senior leaders of the Intelligence Community is essential, but it has never been successful when attempted because the DCI, also being the director of CIA, has not been able to divest himself of the parochial interests of that role. Such a council can provide the DCI, when he gives up his CIA hat, with advice on the wisdom and likely consequences of his decisions before he makes them. It can create a sense of institutional coherence, allowing heads of various Intelligence Community components to participate in decision making and have frequent, direct interaction with the DCI. This last point is extremely important because the interdepartmental nature of the Intelligence Community makes the DCI's executive role difficult. Some sort of cohesive council headed by the DCI can provide a vehicle that helps him impart clearly the rationale and aims of his decisions throughout the entire Intelligence Community. Not only will heads of Intelligence Community components be better able to implement these decisions; they can also better explain and defend them within their own departments. If such a council (today it would be the Intelligence Community Executive Committee) is to work effectively, the CMS must be its staff organ, preparing its agenda and providing the staff analysis to support its deliberations. No less important for the council's success are the recommended changes for the collections disciplines, treated later on, because they will determine whether the members of the council can provide the kinds of resource management support the DCI needs to become more effective.

Recommended Changes for the Position, Role, and Authority of the DCI

Make No Statutory Changes in the DCI's Authority

The DCI's formal statutory authority is adequate. Perhaps some things that are within his authority by virtue of precedent and practice but not necessarily in law or executive order should be codified, but there is no pressing need for it. Two additional authorities for the DCI have occasionally been strongly recommended. It is necessary, therefore, to explain why they are not advisable.

First, greater personnel control has been proposed for the DCI in intelligence organizations beyond the CIA—specifically, the authority to select the directors of the Defense Intelligence Agency (DIA) and the National Security Agency (NSA). The DCI does need to take an active role in personnel policy making throughout the Intelligence Community. For example, personnel security clearances are granted on the basis of widely divergent criteria from agency to agency. Shared standards make sense in this regard, and the DCI should try to achieve that goal. Education and training policies for the Intelligence Community need DCI policy attention. But these issues can be addressed under his current authority.

Appointing senior military officers to lead such agencies as the DIA, the NSA, and the National Imagery and Mapping Agency, however, is not within the DCI's prerogative, nor should it be. These defense agencies are within the military command system. Their directors are chosen from among flag officers by the Joint Chiefs. Seldom does an incumbent DCI personally know the whole set of officers from which a director must be chosen. The Joint Chiefs do. The director of NSA in particular is in a position precisely analogous to all the commanders of unified commands. No one would dare propose that some outside appointed official be allowed to interject himself into the military chain of command to select officers to become commanders in chief of a couple of unified commands. If the DCI had that authority, it would ensure permanent and deep resentment by the Joint Chiefs and unified commanders, making the DCI's management of the Intelligence Community extremely difficult.

Most DCIs, when they have wanted to influence the selection of directors of NSA and DIA, have been able to do so by informally expressing their views to the secretary of defense and the Joint Chiefs. Neither the secretary of defense nor the Joint Chiefs are inclined to appoint an officer who is unacceptable to the DCI. When the DCI's first choice has been refused (as it has on occasion), he has usually been able to exercise an informal veto of candidates he strongly opposes. Given the interagency character of the Intelligence Community, this informal exercise of influence over senior Intelligence Community personnel appointments is much preferable to formal authority.[1]

Second, congressional oversight committees have from time to time proposed giving the DCI budget execution authority (in addition to program management authority) over the National Foreign Intelligence Program. From the oversight committees' viewpoint, this would simplify and strengthen their watch over the execution of the intelligence budgets. At the same time, though, it would complicate their relations with the armed services committees, which are not inclined to share this oversight role. Far more complicated would be DCI management of budget execution inside each of the military services and inside the Department of Defense, Department of State, Department of Justice, and so on for various agencies. Budget execution, by its administrative as well as its management nature, must be done through a direct organizational hierarchy of responsibility and accountability. Having two overlapping budget execution authorities trying to manage jointly the spending of the monies not only would inspire endless bureaucratic turf quarrels, it would make responsibility ambiguous and accountability difficult. Giving budget execution authority to the DCI for every account in the Intelligence Community budget would be a cure far worse than the disease.

Program management, the building of a unified intelligence program to be presented to the Congress for authorization and appropriation, is another matter, and the DCI performs this task for the entire National Foreign Intelligence Program. This authority of the DCI is a powerful tool for deciding allocations throughout the Intelligence Community. Once the budget is adopted by the Congress, of course, agency discretion in budget execution is fairly limited. Thus the agencies of the Intelligence Community are not at liberty to ignore the DCI's resource allocation preferences at will. If the DCI wants to exercise management approval of agencies' reprogramming requests before submitting them to the congressional committees for legal approval, he can do that, using the CMS.

The major program management problem the DCI might face would be a decision by the secretary of defense to reprogram funds within the Department of Defense's portion of the foreign intelligence budget. The secretary's formal fiscal authority is stronger than the DCI's program management authority, allowing him to ignore the DCI's preferences if he chooses. In such cases there is no tidy administrative solution to what are essentially disputes between the DCI and the secretary. If the DCI feels strongly about such a dispute, however, he has a route of appeal. He can ask the Office of Management and Budget (OMB) to take the matter to the president for resolution. This is the DCI's real source of bureaucratic power. Because the budget is, in reality, the president's, handled by the OMB, the president can give the DCI whatever authority he chooses in managing it. It is difficult to see how new or different legal authorities for the DCI could improve the present situation.

The changes in the DCI's status which are most needed have to do with his style and the clarity of his position as director of central intelligence as distinct from director of the CIA. If the CIA is to be retained, the DCI, who has traditionally been closely identified with the CIA and loosely identified with the Intelligence Community, should reverse those relations, preferably yielding the post of director of the CIA to another individual. As part of this strategy, the DCI should maintain clear borders between his offices and personal staff, the CMS, the NIC, and any other special support structures within the CIA. The separation must be real, not merely symbolic.

The temptation to increase the formal authorities of the DCI by statute is understandable although ill-advised. Congressional intelligence oversight committees naturally tend to want the DCI to answer to them on the same basis as any federal agency head, cabinet level or lower. In reality, however, this is not possible because the Intelligence Community is interdepartmental, and it cannot be otherwise. Intelligence, as my discussion of doctrine made clear, is not an independent function that can be wholly autonomous from the organizations it supports. The Intelligence

Community and the DCI's control of it simply have to be viewed as a special case, cutting across organizational lines in an often untidy fashion.

The changes that do make sense in light of the DCI's two major areas of responsibility—1) collection management and intelligence production, and 2) resource management and Intelligence Community policy—can all be made within his present legal authorities.

Strengthen the Role of the National Intelligence Council

The role of the National Intelligence Council (NIC) should be strengthened as the DCI's instrument for a variety of tasks:

- overall collection management in the Intelligence Community;
- providing analysis at the national level that is not produced by any other analysis agency or section;
- overseeing analysis and production in all Intelligence Community components;
- overseeing an Intelligence Community–wide system of data files and materials from all sources and ensuring that these files are kept available to all intelligence analysis units.

Collection management would be a new function for the NIC, but a sensible one, according to the chapter on principles and concepts. It is a staff function that initiates and directs the entire intelligence production cycle. Currently this responsibility resides with the Community Management Staff (CMS), and arguments can be made to retain it there. But because deciding what should be collected, aggregating those choices, and prioritizing them as requirements are so closely related to what intelligence analysts need

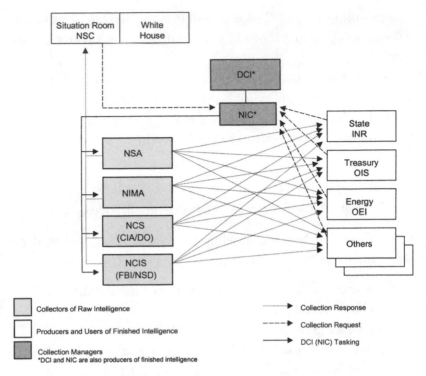

Figure 2. Reformed Collection Management
and Intelligence Production for Civilian Departments

and what they should be doing, NIC personnel are likely to be better informed for carrying out this function effectively than CMS personnel, who are more distant from intelligence production activities. Figure 2 illustrates the reformed NIC's collection management role for the non-time-sensitive raw intelligence needed for intelligence production in the civilian departments.

The NIC, of course, cannot undertake most time-sensitive collection management, either for military operations or, with some exceptions, for political crises. The National Security Agency has a highly effective system for time-sensitive collection management challenges, and to a lesser degree, such a system exists in the Defense Intelligence Agency to support the Joint Chiefs of Staff. Largely these systems deal with signals intelligence collection; imagery intelligence and human intelligence have always been behind in responding to crisis demands for rapid shifts in collection. I will later propose remedies for this situation. Here it should suffice to emphasize that the problem cannot be solved at the NIC level. The NIC should help the DCI ensure that time-sensitive collection management can be handled by each of the intelligence collection disciplines, but the communications, knowledge of collection sources, and the technical nature of actual directives for shifts in collection are complex and different for each discipline. That means that each discipline must solve them separately but within an Intelligence Community–wide concept of collection management operations. Staff supervision in the design and creation of the collection management system does lie within the responsibility of the NIC.

In intelligence production, the DCI cannot be the only voice on intelligence judgments, though he can be the final authority in disputes over national-level judgments. Every department has its own policy or operational issues which, if supplied adequate intelligence analysis, are beyond what the DCI can know about and be directly responsible for. Neither the DCI nor the CIA, for example, can know about or provide the comprehensive technical intelligence needed to support a navy weapons program, or an air force aircraft R&D program, or an army tank program. Even political intelligence needed to support decision making by the National Security Council is seldom the monopoly of the DCI. The president's national security adviser and his staff are frequently better placed and informed for making key intelligence judgments. At times in the 1980s, the secretary of state was far better informed about Soviet leadership views than the DCI or any part of the Intelligence

Community. The secretary simply spent so much time with Soviet officials—and his immediate staff aides interacted with Soviet Politburo members' staff aides—that he had better insight into the top leadership circles. The secretary of defense at times is vastly better informed on foreign military matters by his own intelligence analysts than is the DCI. This is normal and to be expected. In fact, an effective DCI contributes to this reality by ensuring an adequate resource allocation and an effective division of labor among intelligence production capabilities.

The NIC and its staff members, however, can range through the Intelligence Community, staying abreast of the state of analysis and production, its strengths and weaknesses. Specifically they should perform the following functions:

- Identify gaps that are ignored because they seem to fall between agencies. Discover duplication. Identify analysis units that are not receiving relevant raw intelligence. Assess the quality of analysis in various Intelligence Community production units. The NIC cannot do all of these things down to the lowest levels, especially in the military services and unified commands, but at the higher reaches of these organizations it can stay aware of what is being done.
- Advise the DCI, based on the findings from these exploratory activities, on formal allocation of intelligence production responsibilities among all Intelligence Community production organizations. Also recommend intelligence analysis initiatives in neglected areas.
- Establish an Intelligence Community system of data storage, retrieval, and recording of all analysis products and supporting databases, libraries, and so on. These central files, of course, should be made available

to all intelligence production units throughout the Intelligence Community, relieving them of maintaining redundant and inevitably contradictory databases and libraries.

By playing this role, the NIC can become the DCI's instrument for managing intelligence analysis throughout the Intelligence Community. And its acceptance in this role by all parts of the Intelligence Community will depend on the NIC's clear separation from the CIA because other parts of the Intelligence Community will seldom acknowledge the CIA's authority to become directly involved with their work.

Relocate the Directorate of Intelligence

The Directorate of Intelligence (DI) must be separated from the CIA and subordinated to the DCI through the NIC. The directorate should be greatly reduced and the areas of intelligence analysis it performs limited. It should become the DCI's personal analysis arm for selected analysis to fill gaps, highlight anticipated problems, and stimulate follow-up analyses in other Intelligence Community production components.

The DI has become too large and bureaucratic to perform innovative and insightful analysis. This is not its only problem. Two others are equally serious.

First, because the DI is not part of the Department of Defense or the State Department, these departments cannot "order" it to do anything. Military commanders and policymakers in those departments therefore have never depended primarily on DI analysis. They may occasionally find some of its products useful, but in their needs for day-to-day and even long-term intelligence production, they depend on their internal analysis capabilities. DI analysts have long been encouraged to get out and "market" their products, and they have tried. But for the most part these efforts are doomed before they begin because of the DI's organizational independence. DI analysts have expressed great frustration that their products have little impact on policy. Management techniques by the DI will not overcome these difficulties; they are inherent to organizational structure. There is simply no way to provide "one stop" intelligence analysis support to all departments and agencies. Yet the illusion has long been sustained in the DI that it is effectively the single source of valid intelligence analysis support. An analogy with changes in computers is instructive: once microprocessors hit the market, large mainframe central processors began to decline in usefulness. The DI has tried to be the "central processor" for intelligence production, but distributed processing has taken the lion's share of the market.

The DI is also spread too thin. It operates in virtually every area of intelligence analysis—general military, technical military, scientific and technological, economic, political, counterintelligence, and so forth. Yet it is not comprehensive in any of these areas. For example, the DI does analysis of foreign tanks. No U.S. Army tank development program, however, could survive on the DI's tank analysis. The DI's work is simply too eclectic, incomplete, or untimely. The same is true for intelligence support to any navy or air force weapons program. In matters of the services' development of military doctrine, the DI's products on foreign militaries would not even begin to provide sufficient information to satisfy their needs.

In economic intelligence, the DI's products are likewise inadequate. The amount and variety of economic information gathering and analysis in both private and government institutions is vast. The DI has no prospects for competing with them or for providing significant supplementary analysis. When the special trade representative or officials from the Departments of Treasury, State, or Commerce are engaged in economic negotiations, the Intelligence Community can supply helpful support, but most of it is raw intelligence analyzed directly by the negotiators and their staffs. DI analysts are seldom in the loop except when they are detached from the directorate and posted on the negotiators' staffs. Political analysis by the DI generally has the same shortcomings as military and economic analysis. Independent intelligence analysis units in the departments and agencies are better placed to make such analysis relevant, and overt sources—the media, scholarly journals, and so on—are more quickly available and normally better.

The solution to these systemic problems is to change the very purpose of the DI. It should not try to provide single-source support to all users of intelligence analysis; it should recognize the age of distributed processing (see figure 2). Nor should the DI try to compete with the military service intelligence analysts on military topics or with other agencies on economic and political topics. Rather, it should be dramatically scaled back in size and converted to a flexible analysis unit. Its tasks should be, first, to look for problems and issues being neglected by other Intelligence Community components, then to develop this information for the DCI, and finally to pass them off to appropriate Intelligence Community components for sustained and comprehensive analysis, if necessary. The DI should not try to compete with the rest of the Intelligence Community in analysis but rather range across all areas, taking a longer view, a more innovative view, probing and pushing in neglected fields to determine whether indeed they deserve comprehensive attention. If so, the DCI can use such exploratory analysis to persuade and encourage one or more Intelligence Community components to take responsibility for continuing analysis. If the White House and the National Security Council staff are concerned about areas not well addressed by the National Intelligence Council and other intelligence production units, the DCI can use the much smaller and higher-quality unit to address these concerns.

Some areas of analysis require diverse and highly specialized skills, and this streamlined DI should not attempt to maintain a high level of specialization. It should be provided with funds to contract outside research on a temporary, task-by-task basis when its staff skills are inadequate. Through this mechanism, the DI should give the DCI access to the rich and diverse set of university and think-tank centers with expertise in particular areas. Most political and economic issues need little or no classified material for first-rate analysis, and for this reason, virtually all of this work should be done on an unclassified basis.

The DI, with this set of changes, would provide the DCI with the means for selected intelligence support to the White House and key departments and agencies without causing disruptive competition with other analysis units. It would provide some depth under the NIC as well, giving it support beyond the Intelligence Community components, to deal with issues that are not routine, not well accepted within Intelligence Community analysis circles, or otherwise not effectively addressed.

Restructure the Community Management Staff

The Community Management Staff (CMS) should be restructured to facilitate the DCI's exercise of his responsibilities for collection evaluation, resource management, and Intelligence Community policy.

The effectiveness of the CMS, a latecomer in Intelligence Community organizational evolution, has varied with the management style of each DCI. Some DCIs have tried to use it fairly vigorously; others have worked around it or ignored it. The first approach requires that the DCI put significant emphasis on his role as head of the entire Intelligence Community, standing above and to some degree apart from his role as the director of the CIA. When the DCI has fallen back on the CIA for his primary staff support, trying to manage the Intelligence Community as the director of CIA, the CMS has been extremely weak.

The various agencies in the Intelligence Community have been ambivalent toward the CMS. On the one hand, they value the CMS as a source of national collection management guidance so that they can prioritize their own resources and operations effectively. On the other hand, they have resented CMS attempts to get into management issues inside the various agencies.

The consequence of this ambivalence has been the emergence of a large CMS mostly without a clear and sustained mandate, staffed largely by personnel drawn from various parts of the Intelligence Community, making it more a representative "parliament" for the Intelligence Community than a genuine staff. A rethinking of structure and function can remedy most of the deficiencies of the CMS.

First, retain the position of a head of the CMS at the level of a chief of staff, who directs the staff's work, ensuring that it is fully coordinated, timely, and relevant to the DCI's needs for decision making. The CMS cannot and should not give orders to the Intelligence Community. Its purpose is to develop decision options, analyze them in depth, and present them with fully developed pros and cons to the DCI for his decision. Although the CMS must make the DCI aware of dissenting views from all Intelligence Community components, its job is not to resolve disputes or to work out only those alternatives that enjoy full consensus. The CMS must derive its authority among the Intelligence Community components from the high quality of its staff work, its consistent openness to factual information, and its innovative development of solutions and recommendations. If it becomes the prisoner of parochial analysis and bureaucratic paralysis, it has failed.

CMS should consist of five primary staff sections:

• Evaluation Management Section. This section would be responsible for evaluating the responsiveness of collection agencies to the list of national requirements compiled by the NIC from all user departments and agencies government-wide.

Traditionally, the CMS has been responsible both for compiling the national requirements lists and for evaluating how well various intelligence collectors have met the requirements. I have proposed that the task of compiling national requirements, prioritizing them, and issuing them for the DCI as directives to collection agencies be moved to the NIC. Such requirements lists have been standardized for many years, and they are an essential mechanism for all federal departments and agencies to register their long-term intelligence needs and priorities. Collection agencies in the Intelligence Community need them for annual and five-year planning and programming for their activities.

The second task, evaluating collectors' responsiveness to the national requirements lists, should remain with the CMS. Much greater emphasis needs to be placed on evaluation of collection responsiveness by human intelligence, imagery intelligence, and signals intelligence agencies. This can provide the DCI with much better information about the Intelligence Community's outputs, which is vital in making decisions about inputs.

The CMS evaluation management staff section therefore must have sufficient staff to take this responsibility more seriously than has traditionally been the case. And its evaluation efforts must include determining whether intelligence requested by agencies is actually needed and used. There is considerable room for innovation in ways that the evaluation process is conducted. The evaluation management staff section must also maintain close liaison with the new Collection Management section created in the NIC. The latter obviously is well placed to report on both collection responsiveness and actual needs for the information. And the former, by pressing for such information, can help make the NIC's collection managers sensitive to priorities and results. In the current arrangement, the CMS can compile the national requirements lists and neglect evaluation. And the NIC and DI are often frustrated about collection management priorities. The change in responsibilities leaves the CMS no option but to deal with evaluation—something much more important for other CMS functions—while giving the NIC control of collection management—something of key relevance to its other duties.

• Resource Management Section. This section of the CMS must handle the Intelligence Community planning, programming, and budgeting process. The budgeting part of this process, as established under congressional direction, Office of Management and Budget guidance, and so forth, must be retained. What the current system cannot handle effectively is determining the relation between resource inputs and intelligence collection and analysis outputs. In other words, it does not impose a planning, programming, and budgeting system over the line-item budget program. The major reason for this failure is the lack of unified budgets for each of the collection disciplines—human intelligence, imagery intelligence, and signals intelligence. One necessary step in addressing this failure is the establishment of a national manager for each collection discipline with complete program control over resources for that discipline.

The director of the National Security Agency is as close to a "national manager" of a collection discipline as the Intelligence Community now has. Yet about 40 percent of the total budget for signals intelligence is outside his program control, most in the hands of the National Reconnaissance Office (primarily a procurement agency, not an operational intelligence organization). Despite the creation of the National Imagery and Mapping Agency, at least half of the resources for imagery intelligence are managed by the National Reconnaissance Office and elsewhere. Human intelligence has no national manager, for, although the CIA deputy director for operations (DDO) has control over most program resources for human intelligence and total control over clandestine human intelligence operations, that control has been ineffectively exercised over Department of Defense clandestine operations.

The obvious and easy solution to this predicament is to designate the directors of the National Security Agency and the National Imagery and Mapping Agency as the national managers for signals and imagery intelligence, respectively, and to make the director of the Directorate of Operations at the CIA the national manager for human intelligence. But that is the easy part.

If they are to be able to "manage" all of the resources for their respective disciplines, then they must have the program budget control (and, to the degree practical, budget execution control) over all monies allocated to their disciplines. It makes sense to give them this control because they are already responsible for the collection output of their disciplines. They run the collection systems. Through their own staffs and subordinate managers they have access to the technical information that is needed to understand what the effect will be if X dollars are cut from their programs and where it makes most sense to add Y dollars if their programs are increased. It is relatively easy to give the human intelligence resources to the director of the Directorate of Operations at CIA. The problem there is that he has never wanted them and normally objects to assuming such responsibility. Giving the directors of the National Security Agency and the National Imagery and Mapping Agency full program control is the hard part.

This is so because the National Reconnaissance Office (NRO) now controls such large portions of resources for signals and imagery intelligence—specifically, all their money for satellite collection capabilities. Although the NRO is a hybrid organization, belonging in part to the CIA and in part to the air force in the Pentagon, the CIA has always looked upon it as its "lever" against the Pentagon's dominance over signals and imagery intelligence. Because the NRO budget is very large, in the billions of dollars, it has also developed a strong industrial lobby that defends it in Congress because it has been a highly reliable source of funding for the aerospace industry. To put program control over the NRO's budget in the hands of national managers of signals and imagery intelligence, therefore, would confront massive political and bureaucratic opposition. Yet that is precisely what has to be done if the DCI is to introduce the planning, program, and budgeting system in the Intelligence Community. That is, if the DCI is to be able to relate resource inputs to intelligence production outputs.

The defenders of the NRO are frightened by the prospect of seeing the NRO budget subjected to such a rational program process. Having no responsibility for the production of intelligence products for users, the NRO is left with the traditional bureaucratic measure of its performance and power: how much money it can spend—the more the better. Cheap ways to obtain the same intelligence output are unwelcome. As long as the NRO controls large parts of the signals and imagery budgets, no national managers of signals and imagery intelligence can effectively propose such cheaper solutions.

There are two ways to give national managers of signals and imagery intelligence program control over the NRO's budget. First, because the NRO is essentially a collection of contracting officers dealing with the aerospace industry, it could be split, assigning one half to the National Security Agency and the other half to the National Imagery and Mapping Agency. Once that is done, the NRO would no longer have a separate budget to defend in Congress. Money for satellite systems would be requested by NSA and NIMA, and both agencies would be in a position to prepare cost-analysis trade-off studies comparing collections systems in space against systems on the ground in the atmosphere, determining the mixes of space-based and earth-based systems that offer lower costs.

Second, the NRO could be retained as it is but without an independent budget. Rather than go to Congress for money, it would "sell" its procurement services to NSA and NIMA. That would leave the program choices of mixes of earth- and space-based collection systems to the two national managers who would have to ask the Congress for the money. Clearly, they would be in a much better position to show the Congress the more cost-effective alternatives in program mixes, just as they would have already done for the DCI's approval of their program budgets.

Either of these changes will allow a reasonably detached assessment of the mix of space-based and ground-, air-, and sea-based collection assets for something approaching an optimum mix—or a "satisficing" (satisfactory and sufficient) mix, to use Herbert Simon's concept of decision-making when the criteria for a truly optimum solution are not available. For the uninitiated, the inefficiencies created by the NRO's independent budget are truly arcane. Surprisingly, many senior officials within the Intelligence Community are just as much at a loss to understand the budget. Yet it has to be understood, because the money at stake is in the billions of dollars, and the potential savings in the past have also been in the billions. Only by implementing one of the two proposed reforms can the potential savings in the future be known and achieved. No other structural change in the Intelligence Community offers such large gains in efficiency in use of money. Funds for analysis and for human intelligence are peanuts by comparison.

Along with national managers of signals and imagery intelligence, there must be a national manager for human intelligence with program control over all Department of Defense clandestine resources, as well as those of the CIA's Directorate of Operations.

Overt human intelligence resources are a much larger challenge to manage, but a national manager can measurably improve the present situation. He would, for example, be responsible for such things as the Foreign Broadcast Information System, which monitors and translates much of the foreign press and broadcast media in the world, as well as for all defector debriefing units and a number of other assets.

A consolidated counterintelligence program with a national manager is also necessary. He must oversee not only the counterintelligence budgets of the FBI and the CIA but also those of the military services. The reasons for such a manager are the same as for the other four: one person and organization have both the control over resources and the responsibility for intelligence output derived from those resources.

With these changes in structure of the Intelligence Community, the Resource Management Section of the CMS would be in a position to present the DCI with genuine alternative program options based on a planning, programming, and budgeting system approach. With the current Intelligence Community structure, that is impossible. No issue is more critical for an effective improvement of the Intelligence Community than this structural problem because it concerns vastly greater financial outlays than any other issue.

• Science and Technology Section. This section should be small and staffed by no more than a dozen of the best scientific and technical people available in the Intelligence Community. Outside scientists should also be brought in for one to three years' service in fields of special importance. Their first responsibility should be to stay abreast of all the leading-edge scientific and technological developments in the world that could possibly relate to intelligence collection potential. Their second responsibility is to stay aware of all the research and development in every component of the Intelligence Community and the Department of Defense, Department of Energy, and any other agencies conducting advanced research and development programs.

Based on information from both kinds of investigations, this section must make recommendations to the DCI for preventing duplications in research programs and for overcoming reluctance to share technology, a serious problem both in the Intelligence Community and in the Department of Defense.

The Science and Technology Section can also play another important role. As I pointed out in chapter 2, the technical collection agencies, such as NSA, become so deeply committed to particular technologies and approaches that they are averse to shifting investments to newer or very different approaches. The Science and Technology Section, when it finds such a situation, could attempt to persuade the agency to invest in different technical approaches. If it resists, the DCI could authorize the Science and Technology Section to contract with a private firm to conduct a "proof of principle" program in order to determine the validity of its judgment about a new technology. If it is successful, then it could be passed on to the relevant Intelligence Community component and the program terminated. Funding for such experimental research and development should come with a strict "sunset" condition—that is, a date at which it will be terminated no matter what it has accomplished. Otherwise, the Science and Technology Section could soon develop its own bureaucratic interest in such programs, pursuing them beyond any practical promise of successful outcomes.

This mechanism for preventing stagnation in research and development is the antidote to the argument that the perpetuation of the National Reconnaissance Office is the only answer to the problem. In fact, the National Reconnaissance Office began as just such a program, and for more than two decades it made great technical breakthroughs, but it also slowly gave way to bureaucratic sclerosis. Unfortunately, it had no "sunset" clause in its founding charter.

• CI Management Section. This section must keep the DCI abreast of the health of all counterintelligence efforts and provide policy recommendations to the DCI for periodic improvements or needed changes. Among the problems it must face is maintaining adequate interagency counterintelligence awareness, helping the DCI overcome the inherently parochial and mutually suspicious climate that affects any counterintelligence organization.

Another challenge it must meet is bringing a multidiscipline approach to counterintelligence—that is, using not just human intelligence collection, primarily "double agents," but also exploiting signals and imagery intelligence and other technical means in support of counterintelligence analysis. Finally, if counterintelligence organizations remain unconsolidated and fragmented, as they now are, among the FBI, CIA, and the military services, this section will have to devote serious attention to balancing resources among them—that is, to helping the national manager.

• Security Policy Section. Clearance procedures and the granting of clearances vary widely in the Intelligence Community and the government. This section must work on steps to reduce the differences and the added costs wherever possible.

It must advise the DCI on security policies for the Intelligence Community. In this regard, information security, to include computers and communications, is a huge new challenge. While the main concern must be on security in the Intelligence Community, this section must stay abreast of all of the work in the Department of Defense and elsewhere in the government devoted to the defensive side of information warfare. And it may find a role in advising about the kinds of intelligence analysis needed for both defense and offense in the information warfare area to the wider national security community beyond the Intelligence Community.

These five Community Management Staff sections will probably need to be supplemented with an administrative section for CMS housekeeping, but for the main staff responsibilities, they should be adequate to provide the DCI with the kind of information and analysis he needs to manage collection, resources, and Intelligence Community policy issues. Most important, they will allow the DCI to introduce the planning, programming, and budgeting system more effectively in relating Intelligence Community inputs to its outputs.

For anyone familiar with the plethora of committees that have traditionally dominated the Community Management Staff, the rather dramatic character of the changes recommended here will be apparent. Those less familiar can be spared the soporific task of trying to understand the current staff structure.

Retain the National Foreign Intelligence Board and the Intelligence Community Executive Committee

As noted earlier, the DCI must foster "community" within the Intelligence Community. The National Foreign Intelligence Board (NFIB) and the Intelligence Community Executive Committee (IC/EXCOM) are appropriate mechanisms for achieving this climate.

The NFIB has established a generally positive reputation for coordinating agreement and recording disagreement in national intelligence estimates (NIEs) and other national intelligence products. The utility of such products for policymaking is not great, and they have become the focus of a lot of criticism and dispute, especially within the congressional oversight committees. Still, the NIEs provide DCI-validated statements of intelligence judgments on key issues, which are sometimes useful in papers of the Departments of Defense and State when an occasional DCI position is needed. In the Joint Chiefs of Staff, for example, these estimates sometimes provide the language for planning documents.

At the same time, NIEs play another, more important, role. The working groups under leadership of the national intelligence officers, regional and functional specialists within the NIC who produce the NIEs, have representatives from virtually all interested analysis units in the Intelligence Community. They have to meet and consider jointly the available collected intelligence relevant to the questions to be answered in an NIE. This process forces analysts throughout the Intelligence Community to deal with a common evidentiary base. If there were no such process, over time, different analysts would find themselves working from different sets of evidence, not always sharing or even being aware of some evidence. The estimate process has the healthy effect of making analysts communicate and share evidence. If the NIEs performed no other service, they would still be entirely worth the effort.

The Intelligence Community Executive Committee has never been effective largely because it was intended to deal with resource input-output relations. For the reasons I have elaborated in the discussion of the resource management section of the CMS—especially the absence of national managers for each collection discipline with full program control—this committee simply is not capable of establishing even rough estimates of input-output relations for the largest part of the budget for intelligence collection. Only the director of the National Security Agency among its members is even close to having sufficient technical information to be able to speak knowledgeably about the output consequences of alternative inputs of resources, and even his knowledge is limited by the NRO's control of large signals intelligence programs.

With a reconstituted committee, this would change. An IC/EXCOM with national managers of the collection disciplines, analysis, and counterintelligence in its membership would be able to give informed advice to the DCI on budgetary and policy matters. Moreover, if the CMS were reconstituted as recommended above, it would be able to present the IC/EXCOM with meaningful options and supporting analysis.

IC/EXCOM discussions in such circumstances could be enormously productive for the DCI and the Intelligence Community top leadership as well. As the DCI became more assertive in exercising his management role, the IC/EXCOM would be extremely helpful to him by providing the Intelligence Community leaders a forum for expressing advice, consent, and dissent. They could not complain, as they so often have, that they were excluded, that the CIA alone was dictating Intelligence Community decisions, or that the DCI was making his decisions on the basis of highly uninformed or patently parochial analysis.

Require Periodic Structural Review

The DCI should be required to conduct a structural review of the Intelligence Community every five years to ensure that its organization is keeping abreast of new technology and guarding against growing dysfunctions caused by inherent organizational behavior.

Most of the needed changes in the Intelligence Community today arise from changing technology and the current community structures' limits to exploiting it fully. Some also are due to inherent organizational behavior that creates dysfunctions not foreseen several decades ago. The Intelligence Community must deal with a dynamic world in two regards. First, as a developer and user of leading edge technology, it confronts steady and relentless change. Second, as an intelligence organization, it confronts a world of varying and changing targets. To cope with both dynamics, it has to adapt, sometimes rapidly. More than fifty years old, the Intelligence Community has, of course, adapted its structures occasionally. But over the past thirty-five years, except for the creation of the National Imagery and Mapping Agency in 1996, no significant structural changes have been implemented, even though this has been a period of dramatic technological advance. At the same time, the Intelligence Community's largest customer, the Department of Defense, has experienced equally dramatic change, due to new technology, to redesign and revamping of its forces, and to rethinking of its regional missions. The Department of State, also a major user of intelligence, has experienced significant changes itself. Treasury, Commerce, Energy, and other departments have become bigger users of intelligence with changing demands.

The Intelligence Community, therefore, should not have to wait on a major crisis to stimulate reform and structural change. Adaptation might as well be built in as a periodic requirement. This is not to endorse change for change's sake or to make a fetish of it. The plethora of studies calling for Intelligence Community reform is disorienting. Organizational reform can be very wrongheaded, disruptive, and even regressive. Which proposals make sense and which should we treat skeptically?

An example of a misguided rationale for Intelligence Community reform is that the end of the Cold War demands it. That is not at all obvious. The dissolution of the Soviet Union certainly changes the targets for the Intelligence Community and demands a new set of priorities. But if the structure of the Intelligence Community were already highly effective in use of resources, and if it were adapted to exploit the technologies it has deployed over the past two or three decades, there would be no case for reform, notwithstanding the emergence of such different intelligence requirements as drug trafficking, nuclear proliferation, and new forms of terrorism. In that case, the Intelligence Community would need only to adopt a different direction in collection management. Its collection agencies and production units would respond appropriately. The new set of output requirements might demand adjustments to the Intelligence Community components, personnel with different language skills and area knowledge, new sites, new data bases, and so on. These changes, however, would require no structural reform of the Intelligence Community, only a somewhat greater shift in resources than happens normally from year to year in the resource management cycle.

By analogy, if one has a passenger airplane that flies regular commuter flights between city X and city Y, and if all the passenger demand dries up for city Y while demand for nearby city Z is growing, all that is required is a shift in the airplane's schedule, responding to the demand in city Z and cutting trips to city Y. No one would recommend that a basically different airplane is needed to make the shift. On the other hand, if the airplane's fuel consumption had been rising, its maintenance costs going up, and its avionics were increasingly obsolete, a new and different airplane would certainly have to be considered. The Intelligence Community today is closer to the latter situation than the former. Reform requirements are not a function of a changed flight route but rather of overdue maintenance, changing avionics, engine efficiency, and other such structural issues.

Another kind of misguided approach to reform arises from excessive concern with technology at the expense of organizational realities. A particular example is worth citing because some critics of this book will inevitably raise it. The technologists have long been fascinated with what is called cross-cuing—that is, using data collected by either signals intelligence or imagery intelligence to "cue" the collection of the other discipline. This concept makes good sense, and it is widely applied even within the present Intelligence Community structure. The technologists, however, are not satisfied with cross-cuing done by collection managers of two different disciplines talking to each other and cuing each other by organizational routines. Instead, they want to automate the process, to link capabilities of the two disciplines with software applications that do the job with little human intervention.

In principle, the idea sounds attractive, but the current organizational realities and the complexities of processing both disciplines from advanced systems make the idea downright ridiculous for the foreseeable future. No doubt, a stand-alone signals intelligence system, narrowly focused on a particular kind of target, and a stand-alone, narrowly focused imagery intelligence system could be coupled through computers and made to cross-cue each other against an equally narrow set of targets. It could probably be made to work if an autonomous organization were created and dedicated to it. But does it make sense? Suppose the target set suddenly loses its importance, like city Y's loss of passenger traffic in the example just cited? The Intelligence Community would be left with the equivalent of an airplane that could only fly to one city. More likely, the owners of the target set will soon become aware of the system and will devise ways to spoof it or mislead it with changed signatures. An absolute certainty is that the new organization created to make the system work will develop a bureaucratic life of its own, squandering resources long after its fecklessness is established beyond a reasonable doubt. A number of military tactical intelligence programs stand as painful and expensive evidence of the wrong-headedness of letting cross-cuing technology ideas get ahead of organizational capacities and ignoring the target realities.

We must, of course, acknowledge that technology does sometimes demand organizational change, that technological solutions may not surprise us with their effectiveness at times, and that high initial costs should not automatically kill high-risk research and development. But we must take some lessons from past failures to exploit technological leaps. First, organizational reform has to catch up with the technologies already fielded and working well, especially when it is reasonably clear that such reform promises considerably improved efficiency in resource use and in performance. Second, venturesome technologists sometimes adopt highly dysfunctional forms and concepts of operating novel technologies.

Finally, some technology schemes, feasible though they are, simply do not have significant real-world applicability.

For actual cases of all three lessons, one only has to look at the past quarter-century of experience with applying computers, advanced communications, and networking systems to organizations of all sorts. Many initial promises have failed. The sales hype is often misleading. Sometimes the great gains made are not the ones predicted. Despite all the "information superhighway" rhetoric, the print media have not disappeared. Nor have libraries with paper books. And the "paperless" office is the rare exception, not the rule.

The economic marketplace forces most of these wrongheaded concepts to be abandoned sooner or later. In the Intelligence Community, where no market mechanism exists to expose such fallacies, the sensible alternative is a healthy degree of skepticism coupled with periodic organizational reviews and reforms.

These recommendations for the DCI's management structure, though some involve major changes, not only in staff support but also within the components of the Intelligence Community itself, essentially call for removing obstacles that block the longstanding evolutionary development toward a more effective community management system. Introducing an Intelligence Community staff and establishing procedures for compiling national intelligence collection requirements lists about three decades ago was a major step in that process. So, too, was the establishment of national intelligence officers and the National Intelligence Council. The trends toward a truly "community" system have long been clear, but two structural obstacles stand in the way.

The first and more serious is the disconnection between responsibilities for intelligence production outputs and program resource inputs. Senior officials responsible for delivering intelligence analysis in the collection disciplines are not fully in control of program budgeting that provides their resources, especially signals and imagery intelligence collection. They are dependent for a large part of their resources on the National Reconnaissance Office. Yet its senior officials are never held accountable for failures in intelligence production, as are the directors of the National Security Agency and the National Imagery and Mapping Agency. NRO officials are, however, informally pressured to bring more business to the aerospace industry. And they are also encouraged to believe they must limit the Pentagon's formal control of the National Security Agency and the National Imagery and Mapping Agency. Quite naturally, both of these agencies reciprocate the struggle for control.

If money were unlimited, this problem would not be so serious, but because all the components of the Intelligence Community must squeeze their program budgets into a fixed total, money spent by the NRO on satellite capabilities that are not really essential must come from the budgets of all the other components. On occasion, resulting cuts have seriously damaged smaller programs outside the NRO. As long as the DCI has no rational system for identifying this kind of misallocation, and given the strong lobby support in the Congress for the aerospace industry, large inefficiencies will persist in the Intelligence Community's budget.

Changing this perverse set of incentives by creating a set of national managers for the collection disciplines will not ensure great gains in efficiency, but it will certainly make it possible. Today it is impossible. The DCI and the Community Management Staff have no way to judge, even roughly, the effectiveness of resource allocations to the technical collection disciplines, namely signals and imagery intelligence. Yet their budgets consume the overwhelming majority of the National Foreign Intelligence Program.

The second structural obstacle is the DCI's dependence on the Directorate of Intelligence at CIA. It discourages the DCI from giving up his second hat, as director of the CIA, and encourages unhealthy competition among the intelligence analysis elements within the Intelligence Community. The real test for intelligence analysis is the truth about adversaries' capabilities and intentions, not whether it "beats" some other agency's analysis. A residual Directorate of Intelligence separated from the CIA and attached to the DCI's National Intelligence Council could abandon such games and play a highly productive role.

Only by reorganizing the Intelligence Community management structures to remove both of these structural conditions will the DCI be able to address more effectively his three major tasks— production of intelligence, resource management, and providing coherence for the Intelligence Community. It is difficult to see how these changes can be made without the DCI abandoning the role as director of CIA. (The director of the CIA could then be the national manager of human intelligence.) And it should by now be clear that if the DCI gave up his CIA hat but took with him the National Intelligence Council (reinforced by a large part of the CIA's Directorate of Intelligence) and a fundamentally restructured Community Management Staff, he would not lack for a bureaucratic base of his own. He would not become, as many observers fear, equivalent to the "drug tsar," the surgeon general, and other White House "tsars" who have pulpits but no programs. Instead, he would be able to assume the leadership position in the Intelligence Community that has too long been lacking, one probably envisioned by the drafters of the 1947 National Security Act but never yet achieved.

The World of Military Intelligence

The focus of this chapter is a top-down management and structure critique of the intelligence organizational structure in the Department of Defense. My aim here is neither to provide solutions to all Intelligence Community problems nor to take an excessively small-grain look at structural issues. Rather I will try to show how a few key structural and management changes at the top levels in Department of Defense intelligence would provide a more effective allocation of responsibilities and missions. With that rationalization of the top management structure, incumbent senior intelligence officers would have a much improved prospect for continuing the rationalization down the intelligence command levels into the military services and the unified commands. In other words, improving things at the top is the first order of business and also about all that can be done in the initial phases of Defense intelligence reform. If that is done well, the reform is likely to expand downward over time.

First we shall look at Department of Defense intelligence organization and processes for providing intelligence support—that is, the way they work for providing outputs of intelligence. Then we will examine the system of resource management—the inputs of money and personnel for Department of Defense intelligence. In dealing with both intelligence support and program management of resources, excursions beyond the confines of the Pentagon are essential because those organizations and processes inside the Pentagon are inextricably tied to the larger Intelligence Community and to the military services and unified commands. Finally, I will make a series of recommendations, based, like others in the book, on the intelligence doctrine and principles for resource management outlined earlier.

The Current Structure for Intelligence Support

The complexity of Department of Defense intelligence organization can be appreciated by looking at the Defense Intelligence Agency (DIA) table of organization (figure 3). Complexity is also evident from a look at the formal and informal roles of the assistant secretary of defense for C3I (command, control, communications, and intelligence). A similar picture emerges from a look at the disposition of counterintelligence responsibilities. Things are no clearer when one examines the top military staff in the Pentagon—that is, the joint staff under the chairman and the Joint Chiefs of Staff. It includes a primary staff intelligence officer who provides intelligence products to the rest of the joint staff as well as the chairman and the Joint Chiefs. The staff section of this intelligence officer is highly dependent on the Defense Intelligence Agency for support. The staff of the chairman of the Joint Chiefs of Staff has not traditionally included a full staff officer for intelligence; instead, a DIA element has served in that role. Dropping down to the level of the military services, each has a primary staff officer for intelligence. Collectively, they are known as the military service intelligence chiefs.

The best way to begin to sort out this maze is to understand that there are essentially two broad categories of intelligence required for the Pentagon and the military services. The first is intelligence support for "materiel and force development." Because

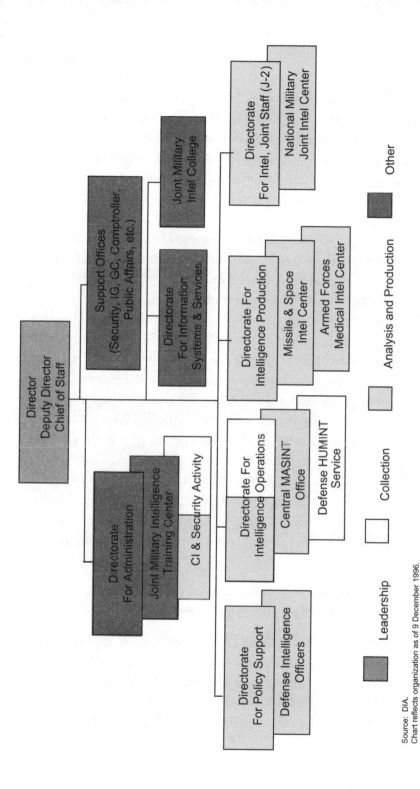

Figure 3. Defense Intelligence Agency Principal Offices, Displayed by Intelligence Function

each of the military services builds weapons and designs forces for combat operations, each needs intelligence about foreign weapons and forces that it might have to defeat in war time. This includes all kinds of technical information about foreign weapons and other materiel, as well as information about how foreign forces are organized and how they can be expected to operate in combat. The second category is intelligence support for daily military operations, war planning, and defense policymaking. The chairman, the Joint Chiefs of Staff, and the worldwide unified commands need this intelligence because they—not the military services—actually conduct operations and planning in both peace and war.

The Defense Intelligence Agency sits astride these two worlds of intelligence, working with the military service intelligence chiefs to provide intelligence needed for the design of U.S. military weapons and forces, and supplying current intelligence to the joint staff's intelligence section as well as to the secretary of defense. We can best find our way through the organizational labyrinth by first examining how the military services relate to DIA in performing their missions and then how DIA and the joint staff's intelligence section perform their missions. After that, we will turn briefly to counterintelligence and a few other defense intelligence issues.[1] That should give us a sufficient picture of defense intelligence to understand how a few changes could improve its performance.

Military Service Intelligence Organizations and the Defense Intelligence Agency

The military services' intelligence organizations are represented in the Pentagon by the principal or senior intelligence staff officers of each military department. These officers, as already mentioned, are effectively the service intelligence chiefs. They have support staffs to assist them in handling their services' intelligence issues for policymaking. They also participate as members of the Military Intelligence Board, chaired by the director of the DIA. They also participate in the National Foreign Intelligence Board, and their representatives are found on most Community Management Staff committees. Thus the service intelligence chiefs stand as the primary management and policymaking link for the DCI and the director of DIA to the services' intelligence organizations and capabilities.

It is also important to understand in general terms what the military service intelligence organizations do. Lower-level tactical intelligence units that are part of combat units deployed under the unified and specified commands are dedicated to providing support to military operations. Each military service has to recruit, train, organize, equip, and field these units. The military intelligence chiefs are not the primary staff officers responsible for these force development tasks. The force structure and program staff sections of the military department staffs manage these tactical intelligence organizations and collection systems. As a result, the Military Intelligence Board, comprising the Director of the DIA, the service intelligence chiefs, and several agency directors, has no program management responsibility over tactical intelligence forces and systems. That responsibility gets lumped into the same staff sections that handle each service's procurement programs and force development issues.[2]

The service intelligence chiefs' most important role concerns intelligence analysis. Although it varies among the services, the service intelligence chief generally supervises a staff that provides intelligence analysis to support all his service's weapons programs and force development issues. In other words, his staff provides complex data of threats against which U.S. weapons and forces must be able to perform successfully. Variations in assessments in threat data can have an enormous effect on the cost of a program.

Thus intelligence support for materiel and force development to the military services is both a large task and one with a major impact on program costs. Throughout most of the Intelligence Community, especially outside the military service intelligence circles, it is not even vaguely understood. At the same time, most of the intelligence collection needed for threat analysis is done outside the military services, mainly by signals intelligence and imagery intelligence, supplemented sometimes by human intelligence (figure 4). The impact on the allocation of large dollar sums caused by this intelligence is real, measurable, and critically important—far more so than for most national-level intelligence products. Yet congressional attention, DCI focus, and the several intelligence reform studies largely ignore the issue of intelligence for materiel and force development in the military services.

How to improve such intelligence and how to make it more effective are beyond the scope of this book, but it is important to be aware of its impact. We also need to realize that the National Intelligence Council and the CIA are essentially out of this intelligence production loop. The science and technology and general military intelligence analysis done by the Directorate of Intelligence (DI) sometimes has an influence on it, but for the most part the CIA's analysts are institutionally disconnected from the materiel and force development processes. The military service intelligence production units and private vendors, performing analysis on contract, are the major providers—a good argument for reducing the size of the CIA/DI and changing its mission considerably, including elimination of its responsibility for production of regular military intelligence. Such products have at most a trivial impact on materiel and force development decisions. If they were not available, they would hardly be missed.

The DIA is dedicated in principle to intelligence for service materiel and force development. It is supposed to manage the

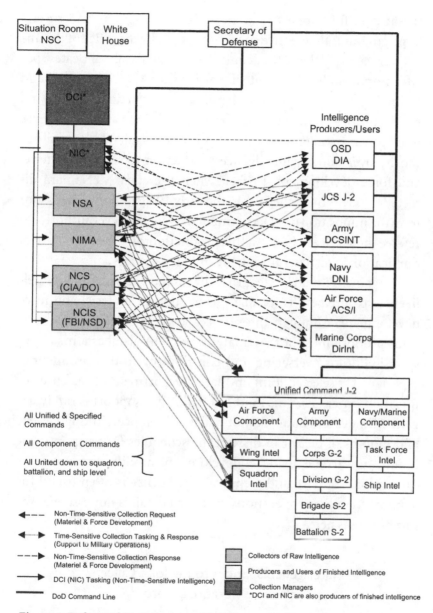

Figure 4. Reformed DoD System of Intelligence Collection and Production for Materiel and Force Development and Support to Military Operations

threat data for support to joint weapons programs and several large programs that receive special oversight by the office of the undersecretary of defense for acquisition. The DIA also allocates specific responsibilities for intelligence production in science and technology among specialized military production centers and holds approval authority over their products as validated defense intelligence. In other words, the DIA holds considerable formal authority over military intelligence production to support materiel and force development. In fact, when Secretary McNamara directed the creation of the DIA in the early 1960s, among his reasons were the imposition of a Department of Defense–level authority to referee the services' efforts in this production, and the avoidance of duplication in analysis among the military services.

Again, it should be clear that the military services' main intelligence requirement is for analysis to support materiel and force development. Although the services' own intelligence organizations produce most of that analysis, the DIA also has a major responsibility for overseeing and contributing to this production, especially for selected joint and large Department of Defense acquisition programs. This kind of intelligence support is rarely urgent. Minutes and hours do not make a difference, although weeks and months of delay can disrupt the schedules for weapons programs. In wartime, materiel development modifications may require rapid support, and technical intelligence is often useful for support to military operations, but such detail is not essential for our purposes here.

Intelligence Support to the Joint Chiefs and the Operational Forces

The Defense Intelligence Agency (DIA) provides some personnel to the intelligence staff section for the Joint Chiefs of Staff. Depart-

ment of Defense intelligence is also linked to the unified and specified commands: the forces deployed under the commanders in chief
with combat contingency responsibilities, primarily through the
primary staff officers for intelligence in those commands. The DIA
maintains intelligence communication links with the joint intelligence staff of the unified commands and some of their intelligence
support units. The joint intelligence staffs are analysis and production staff sections. They may be supplemented by dedicated analysis
units, but each is a commander-in-chief's staff element that directs
collection and produces analysis for his planning and conduct of
operations. Each service component within a unified command
also has its own collection and production capabilities, and they
must be integrated into the command's intelligence system.

Again, recalling the doctrinal principle that distinguished between collection management in general and technical collection
management, let us turn to how intelligence collection should, and
sometimes does, work for the operational forces. The collection
disciplines, of course, are signals intelligence, imagery intelligence,
human intelligence, and counterintelligence. For illustrative purposes, let us take the case of signals intelligence.

Signals intelligence collection capabilities are found in some
of the military units at the tactical and operational levels. And substantial signals intelligence capabilities exist at the national level,
under the direct operation of the National Security Agency (NSA).
At the joint staff level in a unified command, and in the upper
naval, air, and ground unit intelligence staff levels in the service
components, the NSA deploys cryptologic support groups. A cryptologic support group comprises personnel with expert knowledge
in tasking the signals intelligence system to collect specific information and in making sense of signals intelligence products when
they are delivered. But personnel in cryptologic support groups
have neither the competency nor the technical information to di-

rect technical signals intelligence collection management. They depend on the NSA to perform that function.

Because the NSA has operational control over the worldwide signals intelligence system, it can redirect relevant parts of the system to support a single local military operation. In the process, however, it often must cease coverage of targets to meet other requirements imposed by the DCI. In crises and military combat operations, the NSA operations center makes these necessary collection adjustments on a massive scale, bringing to bear a wide array of systems on the appropriate targets. The degree of technical expertise and diversity of skills required for this kind of collections management is beyond what any unified command's intelligence staff can possess. This is also more than the DIA support element with the Joint Staff in the Pentagon can do. Leaving technical signals intelligence collection management to the NSA, therefore, is the only practical alternative. Unfortunately, many military intelligence officers do not understand this reality, and that has caused numerous counterproductive turf battles in the area of signals intelligence collection management.

Because of the peculiar nature of the discipline, effective use of tactical signals intelligence is also frequently better done under the technical direction of the NSA and not by the tactical unit's technical controllers. There are exceptions, but experienced signals intelligence personnel can quickly recognize the situations in which each strategy is appropriate. Tactical-level collection management authorities are unqualified to be involved in technical affairs. Their business is specifying what they need, how soon they need it, and where it is best delivered directly.

The technical aspects of signals intelligence collection are important to emphasize for another reason. Commanders of unified commands and others frequently look at the NSA technical management system as well as other specialized intelligence structures as "stovepipes"—that is, structures within their commands that answer not to them but to commanders outside their command. Stovepipes are viewed negatively, as impediments that should be broken down. But nothing could be more wrong when it concerns "technical collection management," and not just management of signals intelligence collection. Imagery intelligence and clandestine human intelligence also require special "technical" collection management when national assets are working in a coordinated fashion with tactical collection assets that are subordinate to unified commanders. To get the extraordinary advantage of technical management from organizations outside of and far away from a regional operational command, the local commander has to tolerate these seemingly uncontrolled and intrusive stovepipe control arrangements.

The way to solve the problem of stovepipes is not to remove them but to make sure that the intelligence staff section at every command level knows how to obtain access to the stovepipe in order to obtain the data it needs from that collection discipline. The stovepipe structure has actually allowed the signals intelligence system to bring national collection assets to bear on tactical collection requirements. This flexibility in making space-based systems and many other systems work for a tactical commander requires not only enormously skilled technicians but also a large and complex communications system. And it requires a diverse set of processing skills which no military unit could maintain, much less employ. The ability to perform this kind of technical management of all kinds of collection systems is so complex that it can be done only at the national level. It was not easy for the NSA to develop, and it requires constant effort to maintain it. Without it, the old fissure between tactical collection systems and national and other systems would reappear, and signals intelligence support would suffer accordingly. Until commanders and intelligence staff officers understand the imperative for some kinds of stovepipe organizational arrangements in intelligence collection, these people will be an obstacle to the very intelligence support they desperately seek.

A similar approach to the technical management of imagery intelligence is long overdue. A major task for the National Imagery and Mapping Agency will be to create just such a technical collection management stovepipe, making it accessible to staff intelligence sections and other users down to the lowest tactical levels. Under such a structure, the present diverse set of imagery intelligence systems could be concentrated in support of military operations under whatever priority the chairman and the Joint Chiefs dictate.

The nature of technical control required for the collectors of human intelligence differs greatly from that of the other disciplines, but the principle of a national system applies to it as well. Counterintelligence is much like human intelligence in this regard. National-level information about the enemy's intelligence collection capabilities and their targets may not be useful in many cases, but a well-developed national counterintelligence system with coordinating authority over aspects of the military services' counterintelligence assets might prove highly effective.

Military commands have considerable means to collect intelligence beyond the national systems. Not only do they have their own specialized intelligence collection assets; they normally use combat forces themselves to collect intelligence. The collection management of these capabilities is the responsibility of the commands themselves. The only exceptions arise when there is a need to coordinate local collection efforts with national efforts—that is, to allocate the proper division of tasks between tactical and higher-level signals intelligence, imagery intelligence, human intelligence, and counterintelligence capabilities.

Two points in this review deserve special emphasis. First, the national collection disciplines can and must be brought to bear in supporting tactical military operations. And that can be done most effectively by acknowledging the highly technical tasks involved and the requirement for certain types of stovepipes to be created and maintained. Second, most intelligence support to military operations is urgent. Nonurgent intelligence support is also required, especially for peacetime planning by the Joint Staff and the unified commands, but this has not been the problem area. Current operations are the greater challenge and of primary importance. Urgent intelligence support, of course, requires adequate communications. Yet they have been generally lacking in all disciplines except signals intelligence, and even there they are often inadequate for dissemination to some combat units.

Counterintelligence Organization

Counterintelligence organizations and operations in the Department of Defense share no common doctrine. The army maintains units whose sole mission is counterintelligence. Army law enforcement belongs to its Criminal Investigation Division, which has no counterintelligence responsibility. In the air force, counterintelligence is the responsibility of the Office of Special Investigations, which is also the criminal investigations authority for that branch. Likewise, the Naval Investigative Service combines counterintelligence with its law-enforcement function. The operational arms for counterintelligence in the services include a large number of units deployed throughout installations and deployed forces.

Within the Office of the Secretary of Defense, counterintelligence is coordinated by one of the secretary's staff sections. The particular assignment of the duty has changed over time, as has the precise meaning of *coordination*. Working up counterintelligence policies for the secretary of defense to promulgate, getting involved in an occasional complex counterintelligence case, and playing some part in overseeing the counterintelligence budget have at different times been included in the coordination function.

This arrangement leaves military counterintelligence essentially headless within the Department of Defense. Neither an assistant secretary nor a deputy undersecretary has line management authority, although incumbents have occasionally tried to exercise it. Both are staff officers for the secretary, not operators or managers. The absence of a line management element at the level of the Joint Staff or the Office of the Secretary of Defense leaves both cooperation and coordination among the services' counterintelligence operations as voluntary matters. It also means there is little management capacity to impose uniform standards among the services.

We have seen, for example, that the army alone separates law enforcement from counterintelligence. The two do not mix well. The methods for successful counterintelligence work are significantly different from criminal law enforcement methods. Good counterintelligence work is far more expensive per average case resolved. In part this is because the aim of counterintelligence is not just to identify and arrest enemy agents but also to learn about the full extent of an adversary's intelligence capabilities. That may require lengthy surveillance periods to uncover an agent's network of support, to discover other agents, and to assess the methods and capacities of the foreign intelligence service. The tendency of law enforcement agencies, on the other hand, is toward early arrests and prosecution. Law enforcement and counterintelligence also normally involve different sets of methods for investigation. Advanced counterintelligence work requires a multidisciplinary approach combining signals intelligence, imagery intelligence, and human intelligence. And it tends to involve worldwide connections that are seldom found in military criminal cases. Narcotics and terrorism, of course, are exceptions, but they make the point too: law enforcement agencies have not done very well in handling these problems because the perpetrators have much vaster resources and better organization—rather like sovereign governments. More-over, these enterprises are sometimes sponsored by states.

Without a Defense Department management structure over all three military counterintelligence organizations, there is no way to elevate them all to an acceptable professional level. Nor is there a way to coordinate joint operations and to consistently share techniques and data.

Defense Intelligence Collection Capabilities

The Defense Intelligence Agency (DIA) was created primarily as an intelligence analysis and production organization. Today the center of gravity of its activities remains in analysis and production. According to the doctrinal principles, analysis and production are distinctly different functions from intelligence collection, and as a rule, they are best kept organizationally separate. As it is now organized, however, the DIA violates that principle because it has a number of collection capabilities (see figure 3).

First, it manages the defense attaché system, an effective overt human intelligence collection asset. Second, it operates the Defense Human Intelligence Service, which combines the human intelligence capabilities formerly operated by the military services. Third, it has operational control over a number of aerial reconnaissance programs that involve imagery intelligence and special signals collection. Fourth, it has continued to manage the collection and analysis of some radar signals, especially for its Joint Chiefs of Staff support element's warning center. Finally, it manages an odd assortment of overt human intelligence collection efforts—debriefings and interrogation programs, for example. Similar capabilities and programs exist in the military services—for example, the army's interrogation battalions.

This potpourri of collection efforts ranges from reasonably effective to dubious in value. Given that the DIA's major task is analysis and production, the question naturally arises as to whether it needs control over any collection assets.

Department of Defense Personnel Security Programs

Security is not, properly speaking, an intelligence function, but rather a command responsibility. Because security clearances for personnel in all parts of defense intelligence are issued by various security agencies, and because counterintelligence operations are concerned with clearances, the Defense Investigative Service was created to oversee that relation. This service is a centralized Department of Defense organization with overall responsibility for clearances within the department, but the military services have also maintained security clearance systems. Moreover, the National Security Agency, the National Reconnaissance Office, and a few other agencies have their own security clearance procedures. In part these smaller security clearance activities are survivors from the period before the creation of the Defense Investigative Service, but they are also a reflection of the low regard in which the standards of the service are held by its critics. Its task is huge, and its resources have never been abundant. Whether it could recruit and maintain a truly first-rate work force is an open question. It has improved, but its job is thankless, and it has never had strong support from the top leadership in the Department of Defense. Other priorities simply push the Defense Investigative Service to the back of the queue of organizations needing attention.

I make no recommendations on the Defense Investigative Service and other security programs in the Department of Defense but security clearance does deserve serious reform attention.

The Defense Intelligence Resource Management System

The system of resource management for defense intelligence programs leaves a great deal to be desired. A brief description of them
will reveal that managers responsible for intelligence outputs do
not have adequate responsibility for the resource inputs side of a
planning, programming, and budgeting system.

Although the National Security Agency and the National Imagery and Mapping Agency are within the Defense Department,
their program budgets are not a part of the General Defense Intelligence Program (GDIP); nor is the air force's part of the National
Reconnaissance Office's program. The GDIP includes the DIA's
budget, as well as a few parts of the military service intelligence activities performing DIA-delegated work. Its budget is also much
smaller than the budgets of the National Security Agency, the National Imagery and Mapping Agency, and the National Reconnaissance Office.

Counterintelligence programs come under a separate management structure, being combined within the Department of
Defense under the Foreign Counterintelligence Program and included in the NFIP with the counterintelligence part of the FBI's
budget and the counterintelligence part of the CIA's budget.[3]

These programs cost several billion dollars, yet they do not
include all of the intelligence programs in the Department of Defense. A large set of tactical intelligence programs exists with budgets well over $10 billion annually. They fall outside of the DCI's
program management—that is, outside of the National Foreign
Intelligence Program. It may be surprising to learn that a huge part
of expenditures for intelligence lies outside of the Intelligence
Community. But intelligence collection and use is such a major
part of all military operations that one should expect the military
services themselves to have large intelligence capabilities.

More than two decades ago the Congress became interested in how much money is allocated to intelligence outside the DCI's National Foreign Intelligence Program. Drawing the line between what rightly deserves to be included and what should be excluded is not easy. Nor is it easy to ensure compatibility, avoid overlap, and ensure interoperability between the capabilities purchased by national and tactical intelligence budgets, but since that complex problem lies outside the responsibility of the DCI, we shall merely note it here in passing.

The assistant secretary of defense for command, control, communications, and intelligence has staff oversight responsibility for all Department of Defense intelligence programs. At the same time, this assistant secretary is not really a program manager. He is primarily the secretary's eyes and policy adviser on intelligence programs. It is in the nature of the position, however, to try to assert program management control in the absence of a line program management authority over tactical intelligence and related activities.

The reasons are easy to understand. First, the assistant secretary receives pressure from Congress to improve the management of the very large budgets for tactical intelligence capabilities. Second, he is in a position to see the functional fragmentation between various Department of Defense signals intelligence elements. Getting the military services to cooperate and avoid overlap in research and development and procurement of tactical cryptologic items is difficult, and making tactical signals intelligence capabilities fit the capabilities of different elements is even harder. The assistant secretary is in a position to see the conflicts between various signals intelligence programs but lacks the means to solve them.

This assistant secretary has, at times, also had responsibility for combining the military services' counterintelligence programs. As noted above, they are much smaller than other intelligence programs, and their coordination is important; yet within the Office of the Secretary of Defense there is no formal mechanism for that purpose.

Overall, a rather chaotic picture emerges of fragmented program control. No senior intelligence official in the Defense Department is in a position to assess input-output relations in an effective manner. The directors of NSA and NIMA, as explained in the previous chapter, are blocked from doing so by the National Reconnaissance Office's control of so much of the signals and imagery intelligence budgets. The director of DIA cannot do so because he controls a potpourri of disparate programs, most of them for intelligence analysis and production but a few for human intelligence collection and bits of technical intelligence collection.

Can Structural Changes Improve Defense Intelligence?

It can be said that the system seems to work—more or less. But does that mean nothing can be done to make major improvements? The argument that improvements are largely a matter of better management and leadership are unconvincing. The major problems, as I have tried to show, arise in large part from the structural arrangements that violate the principles set forth in the discussions of doctrine for intelligence operations or for resource management. That is why significant improvement is beyond the power of managers to achieve.

The first glaring violation of those principles is the potpourri of activities—analysis, collection, and so forth—lumped into the Defense Intelligence Agency (DIA). The fragmentation of counterintelligence is another. Even the intelligence analysis function, clearly the core of the DIA, is organized incoherently. No clear organizational distinction is drawn between intelligence support to current operations and support to materiel and force development. And in light of the principle of unifying each intelligence collection discipline, there is no justification for the DIA to be in the collection business at all. What it really needs to emphasize is its collection management capabilities for maintaining access to the national collection systems for signals intelligence, imagery intelligence, human intelligence, and counterintelligence. Competing with those systems (where they now exist) is not a productive endeavor.

Clearly, major structural changes could bring improvements in performance, both in intelligence activities and resource management. But the current organizational arrangements and program responsibilities make significant improvements impossible. Figure 5 illustrates the recommendations that follow, as they pertain to the Defense Intelligence Agency.

Reform in Department of Defense intelligence presupposes adoption of the six recommendations made in chapter 3 concerning the DCI's management structure for the Intelligence Community. That would create the context necessary for most significant reforms that can be implemented in the Defense Department's intelligence structure. The most important of those is the creation of national managers for the three intelligence collection disciplines and counterintelligence with program budget responsibility for their disciplines in the National Foreign Intelligence Program. These changes would require that the DIA yield program management over its human intelligence collection capabilities to the director of the CIA's clandestine service—that is, the director of the Directorate of Operations. Likewise, any DIA signals and imagery collection programs (very small in any case) would be shifted to NSA's and NIMA's programs.

• Keep the Defense Human Intelligence Service as a single Department of Defense organization under the operational control of the director of operations of the CIA.

This change makes the director of the CIA (who is now dis-

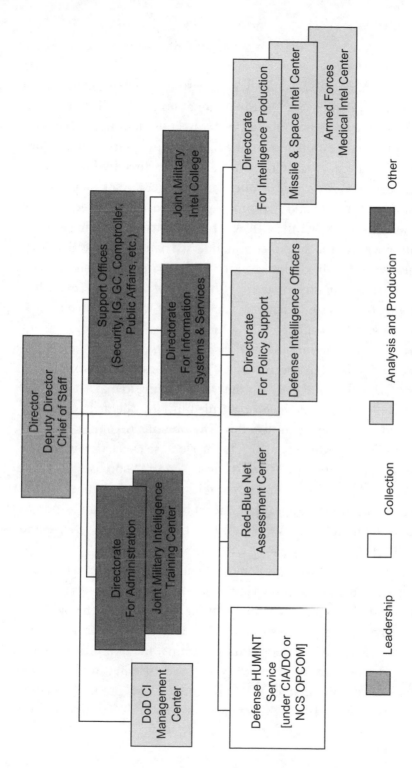

Figure 5. Defense Intelligence Agency after Proposed Reforms. Principal Offices, Displayed by Intelligence Function

tinct from the DCI because double-hatting of one person for both jobs has ended) effectively the national human intelligence manager. At the same time, it does not solve a nettlesome problem for wartime clandestine human intelligence support to military commanders, because the CIA is not under the directive authority of the secretary of defense, as the National Security Agency and the National Imagery and Mapping Agency are. Can the secretary of defense be sure that the CIA will be responsive to intelligence requirements of the unified commanders? This is not an academic issue; it has troubled almost every military operation since 1947. Support to military operations has always been extremely low among the CIA's internal priorities. Thus it was incapable of providing effective support during the Persian Gulf War, and unconfirmed reports of troubled relations between the CIA and the Central Command in Afghanistan suggest that things have not improved. The CIA will be responsive if the DCI is willing to make it so, but if not, only the president can overrule him. Presidents have a poor record of settling such disputes. The obvious organizational answer is to subordinate the CIA to the Defense Department, but the agency's relations with State, its covert actions responsibilities, and several other factors would be complicated by that change. I therefore recommend no specific solution but rather flag the issue as a continuing problem for the president, the secretary of defense, and the DCI.

• Create an overt human intelligence organization within the Department of Defense as a joint activity that coordinates its activities with the CIA as the national human intelligence manager.

This organization should manage the defense attaché system as well as all of the debriefing programs, including operational control over some of the military service debriefing and interrogations capabilities that are not committed directly to unified commands. The model for this organization should be a news service such as Reuters or the Associated Press. Its communication system must allow military attachés, interrogators, debriefers, and other reporting assets to be managed as a network of journalists is directed by a central editorial and production unit. These activities should not only be coordinated with the national human intelligence manager in connection with his responsibilities for a national overt collection system. They should, in many cases, be under his operational control as well. Only then will he be in a position to deal with his overall resource management role for overt human intelligence.

• Put all the Defense Intelligence Agency's electronic intelligence collection under the National Security Agency. Put its imagery intelligence collection under the National Imagery and Mapping Agency (NIMA).

If intelligence analysis units in the Joint Chiefs of Staff or the DIA need electronic intelligence, they should rely on the NSA to provide it. For imagery intelligence, they should rely on NIMA. For measurement and signature intelligence and other special collection, the DCI should assign the collection responsibility to one or more of the collection disciplines. The NSA and NIMA can do some of it. The CIA will also have to deal with some of these requirements. Measurement and signature intelligence and related new technical collection requirements have been a management problem for some time and need the DCI's constant attention as technologies emerge and proliferate.

• Create a Department of Defense counterintelligence management center with operational control and policy, and program management authority over the military service counterintelligence capabilities. This organization should take charge of Department of Defense counterintelligence above the tactical level, and it should relate Defense counterintelligence to the national management system.

• Abolish the National Reconnaissance Office and transfer its program offices to the NSA and NIMA. Alternatively, retain the NRO but give it no independent budget. It would sell its procurement services to NSA and NIMA, which would include funds for satellite procurement in their own budgets.

This change is explained in detail in the sections on DCI management structures, and the signals intelligence and imagery intelligence collection disciplines. It needs no further comment here except to point out the implications for Department of Defense intelligence program management. Major collection agencies should not have large portions of their operating assets funded through an independent Department of Defense program. Signals intelligence, imagery intelligence, and most human intelligence program management will fall under the national managers of these collection disciplines. These national managers will deal directly with the DCI, not through an intermediate line management authority for Department of Defense intelligence programs.

All of the foregoing changes are essential to make the Department of Defense intelligence structure fit the reforms of the Intelligence Community at large. What additional structural changes make sense within the Department of Defense?

• Make the director of the DIA the coordinating manager of all intelligence support to materiel and force development—both joint and by the services.

This organization must perform selected analysis, both general military and in science and technology, and it must keep adequate databases for research and analysis support. But it should allocate, as it now does, the bulk of the analysis to the military services. It must be active in overseeing and encouraging the service analysis efforts, not competing with them. This organization can also be extremely helpful to the DCI if it works with the National Intelligence Council and its director of intelligence in sorting out the more effective divisions of analysis responsibility within Department of Defense and the military services.

• Create within the DIA a "net assessment" center, responsible directly to the secretary of defense, which makes judgments about the strength of U.S. forces versus potential adversary forces as a basis for deciding military force requirements.

Net assessment has traditionally been a point of contention between the CIA and Department of Defense. The Joint Chiefs and the secretary of defense have rightly argued that because they have final responsibility for judging the required size and types of military forces needed to defend the country, they must also have full authority over net assessments. If such assessments were made by the CIA, and if the agency unjustly overvalued U.S. forces or undervalued adversary forces, political groups in Congress and elsewhere would use these assessments to oppose an adequate U.S. military force structure.

Notwithstanding the validity of this argument, the Joint Chiefs and the secretary of defense have never really found a way to do joint net assessments themselves. Each service fears that it could lose force structure or programs as a result of such analysis. Thus the Joint Chiefs agree to disagree by refusing to accede a joint net assessment capability which they cannot fully control.

If the military services cannot agree among themselves, there is nothing to prevent the secretary of defense from creating his own net assessment capability. He could, for example, use DIA resources to create a center within the Department of Defense intelligence structure. The "red" data would be fully available to such a center, and the Program Analysis and Evaluation operation of the Office of the Secretary of Defense could make available adequate "blue" data. The service chiefs might quarrel with the secretary about this organization's analysis, but they could hardly object to his directing that it be done. Moreover, it would be accomplished by Department of Defense intelligence personnel supplemented by people with special skills and training in net assessment techniques and military officers with knowledge of military operations.

In this review of the current state of Department of Defense intelligence I have identified a sufficient number of structural problems to make a compelling case for structural reform. Much of it is necessary to accommodate reforms in the collection disciplines and the DCI's Intelligence Community management structure. Other parts of it are demanded by dysfunctional aspects of the current organization.

The recommendations offered as remedies are imperative where they are necessary to accommodate reforms of the larger Intelligence Community. The other recommendations suggest possible practical applications of the doctrinal principles outlined earlier. Variants might prove to be better solutions, and a few of them might prove difficult to work effectively. The main purpose remains, however, to point the way to reform by offering fairly specific recommendations for structural changes. Getting going— and going in the right direction—is the essential goal.

5

Listening to Learn
Signals Intelligence

O f all the collection disciplines in the Intelligence Community, signals intelligence is the best structured to exploit changing technology and to provide support to both national-level users and tactical military forces. This is true primarily for two reasons. First, in the early 1950s, the military service signals intelligence organizations were centralized under a new organization, the National Security Agency (NSA). The military services' signals intelligence organizations, the Army Security Agency and Naval Security Group in particular, resisted the change and were able to prevent the NSA from gaining program budget control over their tactical assets. They also maintained independent signals intelligence commands, called service cryptologic elements, but the program budgets of these elements were put under NSA management, and their collection assets above the tactical level were placed under the NSA's operational control.

At the same time, however, the NSA's independent budget authority and its autonomous personnel system for recruiting civilians with appropriate technical and other skills created a concentration of resources devoted entirely to signals intelligence under a single director. This allowed a rapid evolution of technology in the

115

NSA's research and development (R&D) programs, permitting the agency to develop systems rapidly. The NSA's central concern with communications gave it a strong impetus to build its own large, flexible communications system. As space communications were added to ground communications, the NSA rapidly built a global collection system. The result has been a fairly dynamic process, prompting constant adaptation in the signals intelligence system over the postwar decades.

Because of its early unification and the nature of its business—technology and communication—the NSA comes close to providing a national manager system for signals intelligence. Among the intelligence collection disciplines it is in the best shape and in least need of major structural change. Still, it has internal problems. It also has one large structural problem that must be overcome before its director can perform an effective role as the national signals intelligence manager.

The Signals Intelligence System

The core of the signals intelligence system is the NSA, which is best understood as a "unified command" within the Department of Defense and also as a "military service." In other words, the NSA is a microcosm of the Department of Defense itself.

Although the NSA's workforce performs the most demanding cryptanalytic work, research and development support, and procurement, the agency also depends heavily on military personnel for most of its collection activities. The service cryptology elements (SCEs) operate the NSA's many field collection sites as well as mobile collection systems. Command of these sites and activities remains with the SCEs, but operational control belongs to the NSA. This distinction, of course, is a product of the doctrine for unified military commands. A unified command normally has command components from army, navy, and air force. These components handle discipline, logistics, personnel, finance, housing, pay, medical services, and all of the traditional command maintenance responsibilities for each service. For the conduct of military operations, however, the joint commander and his staff enjoy operational control: they give the orders and directions for combat operations. If this doctrinal concept of joint military operations is understood and applied to the NSA's relation to the SCEs, then the system of operations is clear. Actual collection and processing of signals is a matter of operational control. The authority for signals intelligence operations remains with the director of the NSA, who is effectively the joint commander.

This approach to signals intelligence operations has allowed a highly effective system to emerge in which most of the field operations are handled by the SCEs while the more complex tasks of organizing and controlling the system are left to the NSA. Relations between the NSA and the SCEs, however, have not been trouble free. At times, the military services have pulled tactical signals intelligence capabilities away from the NSA's operational control. These moves, which have typically arisen from a lack of proper technical understanding on the part of the military service of the signals intelligence system and process, have always occasioned serious disputes. The NSA has also contributed to the disputes, often because many of its civilian specialists are not familiar with the signals intelligence needs of military combat operations.

A number of crises, such as the Iran rescue attempt in 1980, the Libyan raid in 1986, the 1991 Persian Gulf War, the 1999 war against Serbia, and most recently the war in Afghanistan helped push aside some of these turf disputes in favor of effective innovations in tactical signals intelligence support. Some commanders in the military services have come to understand that having their tactical signals intelligence assets coordinated within the overall system is a huge advantage. And the NSA has used its vast communications system to ensure rapid and effective dissemination to tactical forces of signals intelligence products collected by national systems. In other words, through effective central operational control, the NSA has been able to bring to bear the entirety of the signals intelligence system—space-based collectors, collection sites far from the zone of military operations, and cryptanalytic and linguistic skills that the military services cannot afford or maintain—for support to tactical operations.

Progress toward an effective use of the entire system for tactical support, however, is not a simple matter. It requires rapid and repeated innovation, and it also requires cooperation in a mixture of technical skills to work out effective operational control plans and directives. Civilian personnel with deep technical knowledge must cooperate with military personnel with equally deep knowledge of military operations. Tensions are inevitable, but the NSA director's status as a military officer in the Department of Defense chain of command has made progress possible. Like any other commander in the joint system, the director of the NSA has to respond to his commander, the secretary of defense, through the chairman and the Joint Chiefs.

Viewing the NSA as a unified command provides an important but incomplete understanding. The NSA is also analogous to a military service. It recruits and trains its own personnel—the civilian component of the signals intelligence system. It has its own R&D and procurement systems as well as its own logistics system. In these regards, the NSA is very much like each of the military services, though much smaller. As in the case of its operational connections to tactical military units, the NSA has traditionally had turf struggles with the military services in R&D and procurement of tactical signals intelligence systems. Within the Defense Department, these programs are controlled by the services and included in the Tactical Intelligence and Related Activities aggregation, with the joint portion of such funding found in the Defense Cryptologic Program.

Drawing a clear line between the NSA and military services in this complex program area is difficult. Not only is it difficult to designate some elements of these programs as purely signals intelligence because they are mixed into nonintelligence program elements, but it is also impossible to declare all the personnel involved in their operations as belonging only to intelligence duties. These turf problems will probably remain unresolved, to be accommodated and managed, not eliminated.

In one regard, however, the military services and their service cryptologic elements could improve the situation. By insisting on independent R&D for tactical systems, the services deny themselves the vast advantages of R&D management experience that the NSA possesses. Occasionally a military service has simply taken NSA-developed systems and adapted them for tactical use, and to great advantage, both in cost and in the time required to put them to use. Progress has been made on this front, and it should be encouraged. The military services already face such diverse R&D challenges that they cannot possibly compete effectively with the NSA's more narrow focus on signals intelligence systems.[1]

Thus far, this review of the signals intelligence system has dealt only with its support of military operations and its resource issues with the military. The NSA also provides signals intelligence support to the White House, State, the CIA, and many other departments and agencies. The system for this support has generally worked very effectively. Thus it does not need major reform attention. Certain aspects of it, however, are instructive when considering Intelligence Community structural reform.

The NSA is a combat support agency within the Department of Defense. At the same time, a large number of its customers are outside of military circles in purely civilian agencies. And the customers have been generally satisfied with the support they receive. The significance of this record is important. On occasion, members of the congressional oversight committees make an issue of whether or not an intelligence agency is within or outside the military because they assume that those agencies within the Department of Defense will neglect civilian intelligence needs. The NSA's record is compelling evidence for rejecting that assumption.

It is true that concern has been raised within the military services about their cryptologic resources being used to produce intelligence for nonmilitary users. Ill-advised policies have sometimes been imposed to restrict the use of their tactical signals intelligence personnel. Still, these sentiments and resulting actions have never had a noticeable impact on the NSA's support to civilian users. And in some cases, these military policies have been reversed when it became clear that keeping military personnel outside the active signals intelligence system prevents them from maintaining adequate skills for actual military operational support.

The reverse concern, that civilian intelligence collectors will not provide effective support to the military services, is genuine. This has always been a notorious problem in human intelligence and counterintelligence, and the concern also exists, although for different reasons, in imagery intelligence.

Space-Based Collection Systems

The National Reconnaissance Office (NRO) has the responsibility for procuring and fielding space-based signals intelligence collection systems. Several decades ago, when such systems were new and few, the NRO was highly effective in pushing the frontiers for signals intelligence collection from space. But the NRO was and remains largely an R&D and procurement organization, not an intel-

ligence organization. It consists almost entirely of contracting officers and technicians overseeing private-sector vendors who make
the systems. As I have explained earlier but need to reemphasize
here, the NRO is thus analogous to the R&D and procurement
commands within the military services. It has its own budget
which it defends in the Congress and executes independently. No
other agency devoted to procurement in either the Intelligence
Community or the military services has this autonomy. All others
must let their budgets be integrated within a single military service's budget or within an intelligence agency's budget.

To further clarify this point, suppose that the navy systems
command had a budget beyond the control of the chief of naval
operations and his staff. The systems command in this hypotheti-
cal scenario has almost complete control over what it chooses to do
for R&D, what it prefers to procure, and how much it chooses to
spend. Funds are taken from the navy's own budget. The only connection to the navy the systems command acknowledges is to accept technical requirements for defining the performance of the
ships and aircraft it develops and purchases. That is, the navy is allowed to specify the size and speed of combatant ships, the range
and weapons performance for aircraft, and the like. The systems
command, however, retains the discretion to procure bigger or
smaller submarines, surface ships, and aircraft carriers. It might
prefer to build many aircraft carriers and few submarines, or vice
versa. It might prefer to build only cruisers, neglecting destroyers
and minesweepers.

Under this scenario, organizational theory predicts that the
systems command would tend to choose to build those things that
resulted in as large a budget as possible. Actual naval operational
requirements for combat missions would be secondary. The vendors with which the systems command does business would lobby
Congress strongly to defend its autonomy because they would be-

come partners with it in dreaming up more expensive and technically exciting projects. And to the degree that this drive for an increasing budget succeeded, the navy's budget for all other activities
would have to absorb reductions to meet the growth of the systems
command.

Now, let us return to the signals intelligence system and the
NRO. Over time, constellations of collections systems emerged,
and as the NSA learned more about how to use them, alone and in
conjunction with earth-based systems, the approach for the signals
intelligence exploitation of space systems gradually changed. Increasingly, the NSA became capable of looking at complementary
mixes of space-based and ground-based collection systems. In
some cases a far better overall signals intelligence capability could
be obtained by cutting back space programs proposed by the NRO
and using the funds for nonspace systems or for additional space
systems that the NRO preferred not to procure. Disputes arose over
the mix of various space-based systems themselves. From its operational viewpoint, the NSA preferred one mix, the NRO another.
The reasons for the NRO's preference are consistent with the hypo-
thetical example of the naval systems command controlling its
own budget. The NRO has consistently chosen a costlier mix,
which would in turn maximize its budget. When the resulting turf
battles and budget quarrels have reached the congressional oversight committees, committee members have failed to fully understand them. The issue is not a matter of personal relations between
NRO and NSA managers. It derives from the very structure of the
resource flows, the incentives they create, and the demands by signals intelligence users for better support.

The NRO not only competes with NSA for funds within the
National Foreign Intelligence Program; it has expanded beyond
that area. In its Tactical Exploitation of National Capabilities program (TENCAP), the NRO has solicited funds directly from the

military services by promising them direct intelligence from space systems. As experienced signals intelligence personnel could easily point out, most collection requires large amounts of human involvement in processing, and raw signals intelligence collected from space seldom is usable. Thus the very idea of independent TENCAP was based on a faulty premise. Yet years of spending on TENCAP passed before the military services began to realize that they were not getting much for their money. TENCAP was simply a mechanism for the NRO to obtain additional funding. NRO personnel have little or no comprehensive understanding of what is involved in providing usable tactical intelligence. Thus they can honestly push technical schemes that on the surface appear feasible but in reality are not.[2] Limited technical understanding within the military services on such matters qualifies them as easy targets for these schemes.

If the NRO's portion of the national signals intelligence budget were a trivial amount, the problem might be written off as an acceptable bureaucratic overhead cost. But during the 1980s it averaged about 40 percent of the signals intelligence budget. The director of the NSA therefore cannot answer effectively to the DCI regarding the national signals intelligence program. Nearly half of it is kept entirely hidden from him by the NRO.

Because of the amount of money spent on overhead collection systems, this structural problem in the Intelligence Community probably accounts for vastly more waste of financial resources than any other. The potential savings have at times been huge—in one case, $6 billion over five years.

Recommendations

1. Make the director of the NSA the national manager for the signals intelligence program and for operational control and manage-

ment of the entire system. The most important prerequisite for this change is the elimination of the NRO's independent budgetary authority, allowing the national signals intelligence manager to make the R&D and acquisition decisions in his discipline. The NRO's program offices with their expert procurement officers must be retained. They have invaluable expertise and long experience in contracting with the aerospace industry for developing and fielding space systems.

The consequences of implementing these recommendations should be apparent. The NRO program offices would remain with a specialized allocation of programs, not a mix of signals intelligence and imagery intelligence programs as they traditionally have managed. They would no longer depend on the NRO to handle their program budgets and to present them to the congressional oversight committees. Instead, signals intelligence program offices would look to the NSA for that function. The imagery program office would look to the National Imagery and Mapping Agency for that function. Figure 6 illustrates the reassignment of NRO programs to the NSA and NIMA, as well as the organization of national agencies for human intelligence and counterintelligence.

Four critically important advantages will accrue if these changes are implemented. First, the vicious turf fights between NRO program offices will disappear because no single office will be producing both signals intelligence and imagery intelligence systems.

Second, space-based signals intelligence systems can be traded off against other systems, ground, air, sea, and mobile. More effort to use space-based systems in conjunction with earth-based systems will be possible because the opportunities can be considered in the R&D phase, not left to chance after they are deployed. The autonomy of the NRO prevents that today. More innovation in coordinated use of constellations of space systems is also likely.

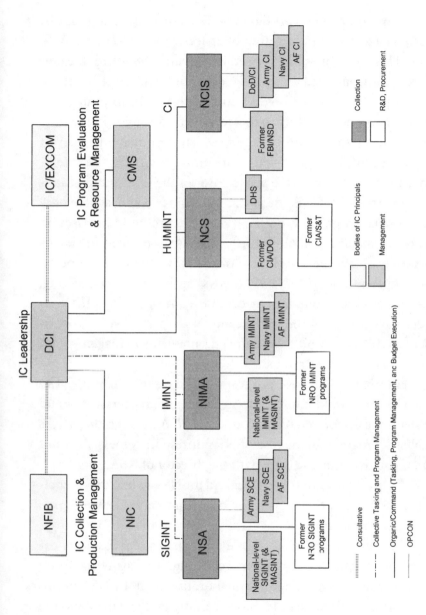

Figure 6. Management and Collection Elements of a Reformed U.S. Intelligence Community Arranged by Function

Third, the endless turf fights between the NRO and the NSA over management and targeting of space-based collectors should abate. This problem was addressed in the early 1980s with the creation of the Combined Overhead Collection Management Center. That step merely papered over many of the problems, but it did give the NSA the capability to redirect overhead collection rapidly at any hour of the day or night. Previously permission to redirect such collection was controlled by a committee under the Intelligence Community Staff, the predecessor of the Community Management Staff (CMS). Because this committee was available only during weekday business hours, it could not respond to crises on weekends and at night. NRO turf prerogatives resulted in ridiculous delays. With the dissolution of the NRO, there would be no question about the NSA's (and NIMA's) authority for retargeting. And the artificial boundary at ground control stations between NRO management of the satellite and NSA (and NIMA) processing of its collection would be removed, enhancing management efficiency within these stations.

Fourth, the TENCAP program would disappear. Support to tactical forces would improve because the military services would not have the option of going around the NSA through the NRO's TENCAP program. In the case of NSA unresponsiveness, the Joint Chiefs could take up the issue with the director of NSA.

2. Direct the military services and the NSA to make greater efforts to coordinate their cryptologic programs and tactical signals intelligence programs.

This recommendation is little more than a plea for more cooperation, but it bears frequent repeating. The army has been weakest in taking advantage of collaboration with the NSA in this regard, and the air force has not been exemplary. The navy and the Marine Corps make the most use of it. A number of joint service programs for platforms that could carry signals intelligence systems have not been fully connected to the cryptologic programs.

3. Direct the DCI to use his new Community Management Staff section for Science and Technology for an examination of several extremely sensitive core capabilities in the NSA. Periodically the NSA has been subjected to an outside investigation of technical experts and scientists to review the health of its strategies, technologies, and other sensitive issues in light of the state of modern communications. Two reasons suggest that such an investigation is needed today.

First, the NSA's telecommunications center ceased to operate for three days in the winter of 2000. Although the agency has taken significant steps to remedy the problem, it is symptomatic of serious technical management lapses as well as a failure to sustain the modernization thrusts it was pursuing in the late 1980s. In the mid- and late 1980s, the NSA was the largest it has ever been, both in personnel and budget. At the same time, it was beginning to address the rapid technical change in the telecommunications world, especially the start of the internet revolution and the shift from analog to digital systems. The end of the Cold War brought severe cuts in money and personnel, playing havoc with most of the modernization programs that had been planned and some that had begun. At the same time, targeting priorities were shifting radically from the Cold War adversaries to many other areas of rising intelligence interest. No government bureaucracy responds very well to sharp cutbacks, and in the arcane world of signals intelligence this was bound to be especially true. In any case, bureaucratic rigidities and the demands of changing are always in tension, and the former clearly began to gain the edge at the NSA in the 1990s.

The second reason for an investigation of the NSA is the growing number of damaging media disclosures about the products and methods of operations of U.S. signals intelligence—encouraged in part by the NSA's new willingness to invite media attention to its activities. A compelling case can be made that if America's first serious signals intelligence organization, launched by the War Department's chief intelligence officer during World War I, had not been terminated in 1929 and its gifted director, H. O. Yardley, not thrown out of work, the War Department would have had early warning of the attack on Pearl Harbor in 1941. Yardley's breaking of the Japanese communications codes was key for the success of U.S. diplomacy at the Washington Naval Conference in 1921. Out of work and in need of income, he wrote an enormously popular book, *The American Black Chamber,* about this episode and dozens of other successes. The book proved a commercial success, not only in the United States but especially in Japan, prompting the Japanese government to change and modernize its codes. One cannot be certain, but had these secrets been kept and had the Black Chamber's operations continued, it is not unreasonable to believe that Yardley would have given the War Department timely warning of the attack on 7 December 1941.

During its five years of post–World War II deliberation on new espionage laws, the Congress considered the probability quite high that Yardley's book contributed to the intelligence failure at Pearl Harbor. The communications intelligence statute of 1951, the toughest law of its kind ever passed, made it a felony to disclose either the substance of any signals intelligence or anything about how such intelligence is collected.[3]

The contemporary American "black chamber" seems to have suffered an analogous defeat, and the appropriate congressional committee might do well to deliberate on the intelligence failure of 11 September 2001 in light of the 1951 law. The NSA lost many important sources during the 1980s due to public disclosures. The most egregious example is James Bamford's book *The Puzzle Palace,* a highly revealing description of the NSA and its capabilities, published in the early 1980s.[4] It became the handbook for hostile intelligence services (in particular the Soviet and Chinese) seeking to penetrate, evade, or otherwise deceive the NSA. This trend of increasing public exposure of its activities and capabilities, encouraged at times by the NSA itself, has surged over the past three or four years. The NSA's failure to detect al Qaeda planning for its attacks on 11 September 2001 raises the question whether this exposure is to blame.

Making the director of the NSA the national manager for all signals intelligence programs in the National Foreign Intelligence Program will create significant opportunities for more effective program decisions by the DCI, but it will not ensure that those opportunities are exploited. The NSA has the technical expertise and the experience essential to clarify input-output relations in signals intelligence program budget, but it never had to do so for the whole of them, especially for the overhead collection systems. The DCI and his Community Management Staff will have to demand that the NSA do this effectively as they impose the planning, programming, and budgeting system on the National Foreign Intelligence Program.

It will also require that the NSA overcome its own internal technical management deficiencies that have accumulated over the past decade or more. Until it has its internal house in better order, it will have great difficulty providing the DCI an accurate picture of the input-output relations for the whole of the signals intelligence system.

Finally, the NSA needs to restore much of its previous anonymity and provide better security for its activities. That will take a few years, but over time memories will weaken, allowing the agency to ease out of the public limelight that damages its sources and productivity.

6

Looking to See
Imagery Intelligence

M aps, drawings, photographs, and a variety of advanced technological means for acquiring and portraying images have only recently been grouped as a single intelligence collection discipline in the U.S. Intelligence Community. Imagery intelligence has yet to acquire the status of a specialty on the same level as human intelligence and signals intelligence. The means for acquiring and transmitting images have involved highly sophisticated technology, requiring great technical skill and competence in processing, interpreting, and disseminating intelligence, but the discipline has received neither a clear definition of its boundaries nor an organizational structure that permits full exploitation of its capabilities.

This can be explained in part as the result of the natural fragmentation of imagery intelligence in earlier times. Drawing was long a standard skill for army officers, especially reconnaissance officers. Sketching terrain, panorama views, fortifications, ports, and the like was one of several standard crafts for engineering officers. Cartography also fell into the engineer's domain.

As photography became available, it was used by reconnaissance officers in wartime, and by military attachés and clandestine agents in peacetime. With the appearance of the airplane in World

War I, aerial photography emerged as a key intelligence collection technique. During and after World War II, the Army Air Corps and later the air force naturally took the lead in this area, but the army and the navy also maintained such capabilities as well.

With new means of imaging—infrared, electro-optics, television, and others—gaining a place in collection activities of the Intelligence Community in more recent decades, imagery intelligence has become a much more important tool. Specialized aerial reconnaissance aircraft (for example, the U-2 and the SR-71) were designed specifically for intelligence collection. The most dramatic breakthrough has been space-based imagery intelligence capabilities. And more recently, unmanned aerial vehicles have been used extensively to gather imagery intelligence.

This rapid expansion of technologies to both provide access to imagery intelligence targets and capture images effectively and efficiently has introduced a complex and rich array of collection means. Creating a coherent and professional imagery intelligence discipline to exploit these capabilities is long overdue. It should be among the major aims of intelligence reform.

The Absence of an Imagery Intelligence System

For years the network of imagery intelligence collection, analysis, and use was too fragmented to deserve the label *system*. Within the air force, however, one could speak of a limited system. Because aerial photography was critical intelligence for targeting bombers, the air force devoted significant resources to aerial reconnaissance. During peacetime, however, the air force could not provide adequate aerial photography to cover the Soviet Union's vast territories in search of missile and other military capabilities. The CIA, using the National Reconnaissance Office (NRO) as its development arm, built the U-2 specifically for the Soviet target, and soon after, it succeeded in establishing imagery intelligence capabilities on platforms in space.

The air force and CIA imagery intelligence programs developed along separate lines. The air force was more concerned with wartime tactical imagery intelligence support. The CIA focused on peacetime assessment of the Soviet military capabilities, especially its inventory of intercontinental ballistic missiles. An overlap between the imagery intelligence interests of the air force and the CIA developed in the Strategic Air Command (SAC). Because SAC's mission was to be able to deliver large nuclear strikes on the Soviet Union, it needed the imagery intelligence products that the CIA was acquiring, but it also needed coverage of more than the military targets of interest to CIA analysts. New and growing space-based imagery intelligence capabilities, launched and largely controlled by the National Reconnaissance Office, were able to meet both user demands.

The navy and the army also began to find uses for this new space-based imagery intelligence. They joined the queue of customers demanding priority for their targets. In the earlier period, imagery intelligence technology for space required a considerable delay from the time of capturing the image to the time it could be exploited by imagery interpreters. In other words, time-sensitive targeting was not possible with space-based imagery intelligence. Air Force tactical imagery intelligence also lagged; photographs had to be taken, returned for development, and then interpreted. These processes could take from a day or so to weeks, depending on the imagery intelligence means.

Because the technology was not initially able to respond to intelligence requests with great speed, the system for its management was constructed with little or no consideration to rapid-response imagery intelligence. The CIA established the National Photographic Interpretation Center to fulfill intelligence requests. The

Defense Intelligence Agency (DIA) and the military services contributed personnel for imagery interpretation to this center, but it remained within the CIA's chain of command. The Committee for Imagery Exploitation, with representatives from all Intelligence Community components, was created to formalize and manage access to the space-based imagery intelligence capabilities. The committee set priorities for all national imagery intelligence systems. Military and civilian users alike queued up to place their requests for imagery.

In principle this made sense as long as imagery intelligence products required several weeks to create. Speed was a major issue only for support to tactical forces, and these advanced national systems were not providing such support. They were supplying imagery intelligence products to the military services, the DIA, the CIA, SAC, and a few civilian departments. Time was not critical for these users. The unified commands in Europe and Korea, however, wanted imagery intelligence that could immediately provide target locations for air and artillery strikes, and they wanted equally rapid poststrike assessments. As weapons systems with increasingly greater ranges began to appear—cruise missiles and short-range ballistic missiles, for example—the need for precise targeting data rose sharply. Attack helicopter units also needed better imagery intelligence to facilitate their deployment deep into enemy territory. Even improved artillery was beginning to reach beyond the army's ability to set targets. Only imagery intelligence was able to provide target identification with sufficient accuracy to support these weapons systems. Signals intelligence could not pinpoint most targets, and human intelligence was too uncertain and often nonexistent.

Several technical developments clashing with organizational realities during the late 1970s and throughout the 1980s account for the contemporary crisis in imagery intelligence. The first set of these concerned intelligence operations. Advances in technology made it possible to produce near-real-time imaging. Although it became technically feasible to deliver such imaging almost instantly to military forces engaged in combat operations, little progress was made in that direction. The NRO—an agency for research and development and procurement—was managing the operations of space-based imagery collection with these new capabilities. The National Photographic Interpretation Center remained the primary place where this imagery was analyzed. Both organizations, under the CIA's control, lacked bureaucratic incentive to rapidly exploit delivery of their imagery products to military forces. Tactical military commanders had no means for tasking the NRO to target its imagery satellites on areas of their intelligence interest. That is, the military had no "collection management" means of drawing on the NRO's capabilities in a rapid and time sensitive fashion the way they could, if they desired, draw on the NSA's capabilities through "cryptologic support groups" that the NSA deploys with tactical military forces.

The second set of clashes between technical developments and organizational realities concerned research, development, and procurement of new imagery collection systems. The military services, especially the army and the air force, were pressing ahead with a radar-imaging capability based on aircraft: the Joint Surveillance Target Attack Radar System. SR-71s and U-2s were still carrying imagery intelligence capabilities. And older air force systems were still active. On the ground, electro-optic systems were being tried in experiments for support to ground force operations. Unmanned aerial vehicles with imaging capabilities were in development. Finally, a revolution occurred in cartography so that map-making was increasingly based on electro-optic imaging from space-based systems. Lacking an overarching program management system, all these technical developments proceeded with little or no attention to duplication, obsolescence of older systems, and operation concepts for making them work together rather than at cross-purposes or with unintended redundancy.

One technical development had a potentially revolutionary impact. As imagery began to be produced in digital form, it could be transmitted by wide-band communications anywhere in the world where equipment capable of receiving it existed. This meant that raw products of imagery intelligence could be transmitted directly to military units deployed for operations. This capability was not immediately recognized. Imagery intelligence products were initially derived by imagery interpreters in verbal form and transmitted as alphanumeric text to users without accompanying photos where speed was desired. As long as this kind of product was acceptable, the National Photographic Interpretation Center and other centrally located imagery intelligence production centers could send their products to tactical forces deployed anywhere. In practice, however, military commanders wanted their own intelligence analysts to have the photos.

The consequence of all these tensions between changing technology and long-standing bureaucratic rigidities was the creation of a high degree of frustration among military users of imagery intelligence. Technological developments had progressed rapidly, but the organizational developments necessary for these new imagery intelligence capabilities to be used effectively for military operations lagged behind. Moreover, there was no organization that could be specifically blamed for this failure. Few if any military commanders or intelligence staff officers knew how the Committee for Imagery Exploitation operated. Some of them looked to the NRO's Tactical Exploitation of National Capabilities program to provide the solution, but it lacked the communications or the personnel to manage the entire imagery intelligence collection and production process. Its managers did not even understand what that would involve.

The organizational vacuum for national-level imagery intelligence capabilities was only part of the problem. An inventory of
the means of acquiring images used in the military services and
throughout the Intelligence Community ranged from hand-held
35mm cameras to satellites with advanced digital imaging technology. Moreover, these many capabilities were distributed in many
different organizations, making collaboration among them impossible without major changes in control and directive authority. No
concept of managing these capabilities as a system existed, and any
effort to create one was certain to face significant bureaucratic resistance.

This was the situation during the Persian Gulf War. Thus it is
understandable why imagery intelligence in that campaign was
judged seriously inadequate. General Schwarzkopf testified to the
Congress on this point, expressing some dissatisfaction with his
generally good intelligence support. Although he did not specify
imagery intelligence as the key problem, that was probably the primary basis for his remark.[1]

The lack of an organization staffed with sufficient skills—
technical and military operational—to take on the task of putting
together a national imagery intelligence system has been a major
failing of the Intelligence Community. The point has been made in
testimony to the congressional oversight committees several times.
Senator Boren, in his draft legislation for Intelligence Community
reform in the late 1980s, recognized the problem and called for a
national imagery intelligence agency. Several other proposals of
this have since appeared.

In 1993 the Department of Defense resorted to something
like the NRO to bring order into the procurement of airborne platforms for carrying imagery (and also some signals intelligence)
systems. Called the Defense Airborne Reconnaissance Office
(DARO), it was charged with oversight of development and pro-

curement of all joint military department and Defense-wide airborne reconnaissance capabilities, including manned and unmanned aerial vehicles, their sensors, data links, data relays, and
ground stations. The hardware programs managed by DARO are
integral components of U.S. national imagery intelligence collection capabilities. But DARO, like the NRO, is a research, development, and procurement office, not an operational intelligence
organization. And like the NRO, it deals with both signals and imagery collection systems, mixing and entangling programs for
each.

Only in October 1996 was the National Imagery and Mapping Agency (NIMA) finally created. This has at least made it possible to introduce an overarching operational control and program
management for all of the imagery intelligence systems. This is a
long-overdue and welcome change as far as it goes, but it still has a
considerable way to go before the director of NIMA can act as the
national manager for imagery intelligence. NIMA incorporates the
National Photographic Interpretation Center (NPIC), the Defense
Mapping Agency, and the ground stations and mission control elements for all space-based imagery intelligence systems. The NPIC
naturally belongs to the core of NIMA. Because imagery intelligence provides most of the raw data for cartography, the Defense
Mapping Agency also belongs within NIMA. The logic of including the ground stations and mission control elements for space-
based imagery intelligence in NIMA should be self-evident. They
are central parts of the present national imagery intelligence capability. Without control over them, technical collection management would be clumsy at best and inevitably delayed when time-
sensitive taskings must be met. Program and budgetary control
over them, however, remains with the NRO—for example, the
CIA's Science and Technology Directorate—and the CIA has consistently struggled to wrestle back what it lost to NIMA in 1996.

NIMA's operational relations with the military services' tactical imagery systems is also apparently tenuous and not effectively institutionalized, not nearly to the degree that the NSA has effective relations with tactical signals intelligence capabilities, which are still far from what they could be. Nor are program budgeting relations sufficiently developed to ensure a well-coordinated national-tactical imagery intelligence system of the kind that technology now makes possible.

Recommendations

The director of NIMA should be designated as the national manager for imagery intelligence. NIMA's structure must allow the director to exercise his responsibilities effectively, in particular, by carrying out the following changes:

- Place the NRO's imagery intelligence space systems development and procurement program offices under NIMA. NRO personnel now run these imagery intelligence programs. The general reason for this step is the same as for giving the director of the National Security Agency control of the signals intelligence programs. As national manager for imagery intelligence, NIMA's director requires final responsibility for the budget now spent on the NRO's imagery intelligence research and development, procurement, and operations programs.
- Assign the primary coordinating and oversight role to NIMA for all military service imagery intelligence programs.
- Direct NIMA to develop a system for exploiting all imagery intelligence collection capabilities to support military operations (or any other operations) in a

time-sensitive manner. This will, of course, require working out coordinated targeting and tasking arrangements with imagery intelligence capabilities within tactical military units.

Program management of imagery intelligence will also remain somewhat divided between NIMA and the military services, just as signals intelligence programs budgeted in Tactical Reconnaissance and Related Activities belong to the services, rather than to the NSA. Many imaging systems are so closely tied to weapons systems and military units that they are beyond NIMA's effective control. NIMA, however, must be highly knowledgeable of all these systems in order to fit over them its array of imagery intelligence capabilities, both those it owns and those under its direct operational control.

Until a system for exploiting all imagery intelligence collection capabilities is developed and practiced, the main advantages of NIMA will not be realized. The foundation for such a system is a worldwide communications structure. On the one hand this communications network must allow central control and direction of imagery intelligence collection and processing. On the other hand, it must provide for rapid and effective dissemination of imagery intelligence products to users. In some cases, those products may only be written reports based on the findings of imagery interpretation. In other cases, they will include direct dissemination to users of the imagery itself, either with accompanying interpretation reports or left to local imagery interpreters to process. Regional interpretation centers, located with unified commands, may prove the most effective solution.

There is no way to know in advance how best to structure such an operational system. Only trial and error combined with experience from actual military operations will allow NIMA to perfect such a system. A key feature of this learning process will be working out methods that allow imagery intelligence personnel with deep knowledge of the collection systems to design plans for rapid selection and exploitation of the best available system to meet urgent intelligence requests. Knowing all available systems, the NIMA operations center should be in the best position to receive, prioritize (under guidance of the Joint Chiefs of Staff), and respond to imagery intelligence requests. It may choose an aerial imagery intelligence platform, a space-based platform, or some other means to get the required imagery. And it must know how best to get the imagery processed into usable form and routed most directly to the user.

Although signals intelligence operations are quite different from imagery intelligence operations, there are parallels, and NIMA could take a number of lessons in this regard for handling crises and military operations where collection assets must be rapidly redirected, communications quickly reorganized, and processing accomplished in an ad hoc fashion.

As NIMA works out its concepts of operations and designates tactical imagery intelligence systems to be placed under its operational control, the secretary of defense and the Joint Chiefs will have to be strongly supportive. Bureaucratic turf issues will be serious, and if they are not addressed by the top levels of the Defense Department, they will place severe limits on the kinds of improvements in imagery intelligence support that NIMA can provide.

Creating an effective NIMA will not be easy. Many bureaucratic interests will be threatened. Elements from several parts of the Intelligence Community will be incorporated, and each will bring along its old ways of doing business. The process will require several years and a series of adjustments. But the potential gains from creating an effective NIMA are simply too great to let the problems stand in the way.

The heart of an effective NIMA is to be found in two major features. The first is its communications system for managing both real-time collection and real-time distribution of imagery intelligence. The second is its management of imagery intelligence programs. If the director of NIMA is given overall responsibility for imagery intelligence, he will be in a position to make trade-offs among various systems, deciding the best mix of space, aerial, ground, and other collection platforms and also deciding among the types of imagery intelligence technologies. Doing these things will allow him to provide the DCI with strong rational arguments about where to cut and where to add to imagery programs. As NIMA now exists, its director cannot provide that kind of program budgeting analysis to the DCI and his Community Management Staff. And without a national imagery intelligence manager, the DCI has little prospect for imposing the doctrinal viewpoint central to my reform recommendations—a planning, programming, and budgeting system—on the imagery intelligence portion of the National Foreign Intelligence Program.

7

Spying to Know
Human Intelligence

The heart of the human intelligence discipline is the clandestine service. A professional peacetime clandestine service is relatively new in American history. Perhaps the most effective American clandestine operations were conducted by George Washington during the Revolutionary War, but operations were discontinued after the war. During the Civil War, Allan Pinkerton's detective service ran clandestine operations on contract in support of Lincoln's administration. Again, clandestine human intelligence was discontinued after the war. In the 1880s both the War Department and the Navy Department initiated modest overt human intelligence collection by military attachés and officers on leave visiting foreign countries. Shortly before and during World War I, some clandestine operations were directed in Mexico because German intelligence was using Mexico as a base for its human intelligence operations. In the interwar period, military clandestine operations again largely ceased. Military attachés began more rigorous overt human intelligence activities, but their output was modest and their judgments of both political and military affairs abroad were often naive. At the same time, the Federal Bureau of Investigation under J. Edgar Hoover's leadership began to engage in foreign clandestine operations in Latin America.

Thus the United States entered World War II without an effective clandestine human intelligence capability. The story of Brig. Gen. William Donovan's creation of the Office of Strategic Services (OSS) is well known. By the end of the war it was extensive, especially in Europe, and it played a critical role in the late 1940s in undercutting Stalin's covert operations aimed at Communist Party takeovers in France, Italy, and Germany.

But Donovan's OSS in many regards focused more on covert action and paramilitary activities during the war than on purely clandestine human intelligence collection. As part of the War Department, the OSS moved into a vacuum. The army's intelligence capabilities were negligible at the time. Donovan engineered enormous advances in human intelligence and covert action, asserting a large degree of autonomy within the War Department. At the same time, the army created the Counterintelligence Corps, which became reasonably effective, and it continued to play a key role in the occupation of Germany and the de-Nazification program.

Donovan always viewed himself as working for the president. He was able to recruit people into the OSS who would not have considered a career in army intelligence after the war. He therefore resisted the idea of leaving the OSS in the War Department after the war. Failing to persuade President Truman, Donovan resigned. The OSS was abolished, but its parts survived as the Central Intelligence Group to become an independent organization, the Central Intelligence Agency.

Although the clandestine service was the core of the CIA, Donovan's successors made the case for a special CIA intelligence analysis element as well. They argued that the failure of army and navy intelligence to warn of the attack on Pearl Harbor proved the need for a national-level independent capability for analysis. These arguments prevailed, and the CIA's Directorate of Intelligence was added. The new director of central intelligence also took William L.

Langer's research and analysis element from the State Department, where it had discovered during World War II that it was not welcome, and made the Directorate of Intelligence in the new Central Intelligence Agency.

The Postwar Period

The military did not wholly abandon clandestine human intelligence after the war. The army's Counterintelligence Corps ran clandestine operations. The air force and the navy also instituted rather modest clandestine human intelligence efforts. By the late 1970s, after a series of organizational changes in the military service intelligence organizations, the bulk of the Defense Department's clandestine human intelligence was in the army. The navy dropped out entirely, and the air force effort was extremely modest. The army, however, had a fairly large and widely deployed human intelligence effort. All three military services retained active counterintelligence clandestine operations.

The CIA Directorate of Operations (DO) became the dominant part of U.S. clandestine human intelligence. Its relations with the army's clandestine human intelligence varied from cooperative to competitive. Holding final approval authority over any military clandestine operation, the DO was effectively in control of the army's capabilities.

For years, the quality of army clandestine human intelligence was mixed. At times and in some places it was ineffective. The CIA worked assiduously to exaggerate the ineffectiveness of army human intelligence. Even army leadership came to doubt its worth. In the early 1980s the chief of staff of the army initiated a study to determine whether or not to abolish army clandestine human intelligence outright, or, if not, to turn it over to the Defense Intelligence Agency (DIA) as a joint Defense Department organization. The study results were not implemented, but a move to create a DIA clandestine effort began to take shape, and in 1995 the Defense Human Intelligence Service (DHS) was formed. The DHS consolidates Department of Defense human intelligence under the DIA.

The quality of several army human intelligence operations improved markedly in the early 1980s. Although the Directorate of Operations was never eager about giving approval authority, a number of operations turned out to be successes. The army made several attempts to work out a division of labor for clandestine operations, but they were never taken seriously by the DO. Part of the reason is to be found in the experiences of the Iran rescue attempt.

The army needed intelligence on the embassy compound in Teheran, and it needed clandestine assets to provide assistance during the operation. The Directorate of Operations had no agents to provide this support and proved unable to develop them during the months of planning for the rescue attempt. The army acted to fill the gap by creating the Intelligence Support Activity (ISA). This unit consisted of a mix of clandestine case officers and Special Forces personnel. A small number of them were able to pass as non-Americans and actually visit Teheran with some DO technical support. After the hostages were released, the chief of staff of the army decided to retain ISA as a permanent organization to support any future operations of this type.

This did not please the CIA. The saga of ISA's efforts to develop capabilities and the obstruction of these efforts by the Directorate of Operations contributed to growing mistrust. By the late 1980s a number of measures had been taken to try to improve the climate. The most conspicuous was the assignment of a general officer to the DO. He was to be responsible for seeing that clandestine human intelligence assets for support to military operations were recruited and maintained by the DO. In reality, although relations between the DO and Defense Human Intelligence Services have improved somewhat, this is not an effective solution. The military officer has no real power to cause clandestine case officers to respond to specific military needs, such as those in Iran and Iraq in 1980 and 1990, respectively.

That the DO would not be very attentive to such developments is understandable. The directorate has never found much value in maintaining networks of low-level assets. Rather it has always seen its mission as penetration of the highest levels of political and military authority in target countries. The pressure to keep this focus was all the greater when the DO was periodically criticized by the congressional committees and others for its shortcomings in this kind of human intelligence access. Neither the DO's self-image nor its ability to impress the president could be enhanced by the kinds of clandestine assets that are periodically needed to support military operations—for example, in Afghanistan in the fall of 2001, when army Special Forces teams needed agents among the anti-Taliban forces. Fortunately, Russian intelligence services came to the DO's rescue by providing a number of contacts in Afghanistan, allowing the DO and the DCI to create the public impression that they performed brilliantly rather than escaping embarrassment through help from their old adversaries in Russia.

Parallel to these troubled relations between the DO and military human intelligence, the directorate maintained highly collaborative relations with army, navy, marine, and air force special operations forces. In particular, Army Special Forces provided considerable resources and capabilities to the DO. In fact, the collaboration became so close that army leadership supervision of it began to break down. In the mid-1980s, when army special operations personnel were discovered operating outside army regulations, it was also realized that this happened in part because the DO had cultivated a sense of separateness and autonomy among the army personnel with whom it had been working, encouraging them not to be constrained by army regulations. Not surprisingly, such behavior by the DO does not enhance its reputation with the army's senior leadership. In the Pentagon's special operations community, however, links with the DO have remained strong for some understandable reasons, but not for intelligence collection. On the contrary, the affinity is a shared interest in covert action, which, as defined in chapter 2, is not actually an "intelligence" activity.

The army's Special Forces have always been in search of missions, and many peacetime covert action programs have needed the very skills they possess. The DCI and CIA, of course, have full and unchallenged authority for conducting all such operations. The Special Forces can participate only at the CIA's request. Most secretaries of defense, however, have been unenthusiastic about Special Forces participation because of the history of covert operations that have become huge public relations problems, and the connection of those operations to foreign policy disasters. Consequently, the CIA has recruited and managed its own paramilitary capabilities. The attitude of the senior leadership in the Defense Department has, of course, markedly changed as a result of the spectacular performance of Special Forces teams in Afghanistan, so much so that the secretary of defense may actually permanently reverse the department's disdain for paramilitary covert action. What will become of CIA's paramilitary personnel in that event, therefore, is an important question. They could be retained to compete with the Pentagon's special operations forces—not a desirable outcome—or they could be greatly reduced, or disbanded, as the CIA begins to depend primarily on the Pentagon for such capabilities.

Two points are key in summing up relations between the Directorate of Operations and Department of Defense human intelligence. First, the DO in principle and authority is well-placed to be

the national manager for human intelligence. The military clandestine human intelligence budget falls within the National Foreign Intelligence Program, and the director of the DO has final approval authority over any military clandestine operation. In other
words, he has the same kind of operational control over military
human intelligence that the director of the National Security
Agency has over military signals intelligence operations. The DO
could, therefore, take the view that military clandestine capabilities
are part of the national human intelligence system and become
deeply involved in their targeting and exploitation. That is the view
that the National Security Agency takes of the service cryptologic
elements. The nature of clandestine operations, of course, is quite
different from signals intelligence or imagery intelligence. Cooperation does not come naturally in clandestine operations. Still, the
CIA's director of its DO has all the authority he needs to take charge
of military clandestine human intelligence. The sizable resources
devoted to clandestine operations by the Defense Human Intelligence Service (DHS) could provide a dramatic enlargement of the
DO's overall capabilities. Moreover, the DO's traditional disdain
for Defense clandestine intelligence capabilities is not based entirely on reality. Even where it is, the DO has the power to help solve
it: DHS case officers are trained in the DO's courses for clandestine
tradecraft. DO instructors are responsible for giving passing grades
only to those military students who earn them.

Second, two separate paramilitary organizations have grown
up, one in the Directorate of Operations, the other in the army's
Special Forces and other Department of Defense special operations forces. The military's strength is in military skills and operations. These are not the strength of the DO. Not surprisingly, DO
paramilitary operations have generally been looked on by army officers as amateurish at best, usually designed to fail. A close com-
parative look at the record of army and DHS clandestine opera-

tions and DO paramilitary operations would, in all likelihood, show that the Department of Defense has more justification for disdaining DO paramilitary capabilities than the DO has for disdaining DHS case officers. This raises the question of whether or not the DO should drop its paramilitary capabilities and depend largely on those in the Department of Defense. Again, the experience with Special Forces teams in Afghanistan seems to have convinced the secretary of defense that such a change is now desirable.

The CIA's DO, of course, does not have its main focus on the Department of Defense. Its primary partner, especially in peacetime, is the Department of State. Because the DO maintains stations in many U.S. embassies, the State-DO connection has always been intimate. Most of the time it has also been cooperative and mutually advantageous, but exceptional cases have made their way into the media at times, and tense and counterproductive relations between the ambassador and the CIA chief of station (COS) are not uncommon. Do these troublesome cases indicate a systemic or structural problem in the management and conduct of human intelligence operations?

The answer is no, but with a caveat. The natural tendency for any COS and his case officers will be to limit the embassy's knowledge of its operations. Lives as well as operational results may be at stake. A prying and skeptical ambassador can easily provoke the COS to leave him less than fully informed. Likewise, a COS can easily provoke deep distrust in the ambassador's mind by being less than forthcoming, causing the ambassador properly to pry into local human intelligence operations. When these operations involve covert action, the chance of State-CIA conflict increases dramatically. When clandestine operations concern only recruitment of spies, it is unlikely to escalate into a major problem. There is no organizational solution to this general problem. It is a management issue both for State and the CIA. The problem can and has, however, taken on larger proportions when it links back to Washington and the DCI-CIA role in conducting foreign policy, and in those cases covert action is almost always the root of the issue.

Most cases in which ambassadors and CIA station chiefs have been at odds to a degree that the matter has become public have their origins in differences between the secretary of state and the DCI. The president has also been involved, generally encouraging or condoning the DCI's actions. This pattern is evident in U.S. policy in Central America in the 1980s, and also in Kennedy administration policy toward Vietnam, although the dynamics were different. President Reagan was more clearly supportive of the DCI and his covert action programs in Central America, while President Kennedy appears to have lacked the leadership insights and experience to control ambitious subordinates, not just in the CIA but also in Defense, State, and elsewhere.[1]

Again, solutions to these problems cannot be found in structural changes in the Intelligence Community. They are policy problems, and they can easily be eliminated by rejecting covert operations as a policy instrument. With the end of the Cold War, the inclination of presidents to resort to covert actions has already declined, but it will not entirely disappear. A number of challenges confronting U.S. policy in several regions of the world will continue to make covert action tempting, though not necessarily successful. The Clinton administration's efforts at covert action against Saddam Hussein in Iraq turned out disastrously and were discontinued.

Finally, a facet of the DO's relation with the State Department deserves a passing comment because it has attracted a lot of public discussion: the DO's dependence on U.S. embassies as its primary bases of clandestine operations abroad. Embassy-based case officers, it is argued, will never do well in penetrating tough targets like terrorist organizations or making recruitments in highly authoritarian states, the so-called "hard targets." An alternative is to have more case officers living clandestine lives under what is known as nonofficial cover—that is, appearing in any number of roles, such as businessmen, doctors, scholars, writers, not using U.S. embassies at all in their operations. This issue should not concern us here because it is not really an Intelligence Community structural problem. It concerns clandestine operational techniques and does not belong to matters of Intelligence Community management and structure.

A related issue concerning the use of embassies as bases of operations is structural in nature. The expansion of the FBI abroad—specifically, the growing number of legal attachés ("legatts") overseas—puts two clandestine organizations in one embassy, and neither likes the other nor wants to cooperate. In the future, this development is likely to become a source of numerous problems.

I shall deal more fully with relations between the CIA/DO and the FBI in the section on counterintelligence. Suffice it to say that the relationship has always been troubled. For several reasons this relationship deserves structural attention. First, in light of changed world circumstances and threats, especially since 11 September 2001, it is increasingly likely that the FBI and CIA will collide in their activities overseas. A better understanding of their competing roles, responsibilities, and missions, and the structures that reflect them will therefore be important. Second, the counterintelligence relationship between the two agencies has long been cloudy. The FBI reacted in what can only be called a celebratory manner to the discovery in the early 1990s that the KGB had recruited a high-ranking DO officer, Aldrich Ames. In fact, the Ames case was just as damning to the FBI as to the CIA because the FBI has the responsibility for surveillance of foreign intelligence operatives in the United States, not the CIA. How had Ames met with his KGB control officers without the FBI noticing? The Bureau was conspicuously silent on this point.

Problems Inherent in Clandestine
Human Intelligence Organizations

Leadership and management of clandestine services confront several unique problems. To some degree, a few of these problems are found in signals intelligence and imagery intelligence organizations, but with human intelligence, including counterintelligence, they take on special features. They present challenges to the management hierarchy that are not found in most bureaucracies, in the government or in the private sector. When members of the congressional oversight committees have spoken of a need to change the "culture" in the CIA, they are talking about precisely these kinds of problems, although they usually have difficulty in being specific about their causes. But they rightly perceive that there is something special about the nature of the problems. *Culture,* precisely because of its vagueness, happens to be a useful categorization. In seeking successful remedies, however, it is essential to overcome that vagueness and specify more precisely the problems as well as their causes. Identifying a few key ones will help make the point.

First, deception and misrepresentation are the heart of clandestine human intelligence skills. Effective case officers must excel in the business of making appearances conceal realities. This is critical for their handling of agents; it is no less critical in techniques for recruiting agents. To put it colloquially, being a skillful con artist is extremely helpful in recruiting and handling agents. The same abilities are not helpful to managers and leaders in hierarchical organizations. If lower- and midlevel managers in bureaucratic structures have a proclivity for con games while managing the reporting of the cold, hard facts of operations both up and down the chain of command, that tendency undermines top management's ability to control and direct the organization effectively toward its goals.

All intelligence organizations (and most military organizations), because of their secrecy requirements, confront this management problem, but clandestine human intelligence organizations are most vulnerable. Skills honed for recruiting and handling agents are easily turned to dealing with undesired management pressures from above. A case officer can easily conclude that higher-level managers simply do not understand his problems in a particular case. He can easily rationalize that if the manager actually did understand, he would agree with the case officer. That means, of course, that a little deception in dealing with the superior is actually in the best interest of the superior. The complex human dimensions of handling agents encourage such reasoning. An agent, when he is recruited, is essentially placing his life in the case officer's hands. The case officer knows this and promises complete loyalty. When telling the unvarnished truth to management might risk the agent's well-being, or even his life, the decision to divulge the whole truth is not easy. The moral dilemmas are painful.

The problem does not stop at this level. As the case officer's career progresses, he may be promoted to a management position. There he stands between case officers and higher management. The same inclinations to play games with the facts come into play. The problem repeats itself right to the top of any clandestine human intelligence organization. Moreover, as time passes, such behavior creates an accumulation of distortion in what the top management knows about the operational levels.

Most experienced clandestine human intelligence managers are familiar with these problems, especially of case officers becoming too attached to their agents and losing objectivity in judging their reliability and productivity. Still, the pressure exists to recruit agents, and the willingness to drop nonproductive agents is mitigated. That climate also accumulates and undercuts management's control over operations.

A second kind of problem is related to the first. The life of a case officer is stressful and anonymous. When he has a dramatic success, he cannot be rewarded by public recognition. Even his family cannot know. After ten or twenty years in the clandestine service, he is a gray and inconspicuous member of any community he lives in, and his children must see him that way. Suppressing one's ego, as this career requires, is not easy. One is recognized for solid accomplishments only by one's fellow clandestine officers. This climate creates a strong bonding effect. Loyalty to one's fellow officers becomes an overriding value. In principle, such cohesiveness contributes to operational effectiveness and useful collection, but in practice it can put the obligations of the clandestine service to the Intelligence Community and the national security interest at odds with the informal group interests of clandestine officers.

The clandestine service is not altogether unique in this regard. Signals intelligence organizations, imagery intelligence organizations, and even analysis units must labor in anonymity to the outside world. They receive recognition primarily from their fellow workers. An indication of this social problem is reflected in a comment by several people about an individual who had performed an incredible feat in a technical intelligence collection operation. They called him "Top Secret Famous." That is, only a few people with top secret clearances in his department knew the extent of his achievement. Coping with the need for recognition while retaining the bonds of secrecy inevitably creates an informal internal cohesiveness that works against admitting mistakes and keeping the higher levels of management fully informed.

In human intelligence organizations, however, the problem is more acute. The personnel often work alone, away from peers who know their worth. And the skills that make them successful operators can easily be turned against their management. Coming from an operational background themselves, the managers are inclined to be forgiving.

A third kind of problem arises from the tensions between counterintelligence and offensive clandestine operations. The case officer wants his agent to succeed; he naturally wants to believe him. The counterintelligence officer is naturally distrustful of every agent. If the ethos of counterintelligence becomes too dominant, offensive operations will suffer. And if the ethos of the primacy of offensive operations becomes too dominant, security will suffer. The history of the Directorate of Operations has been the swing from one kind of dominance to the other.

Fourth, a clandestine service suffers a setback beyond the obvious if it suffers a high-level penetration by a hostile intelligence service. Once the penetration has been discovered and publicly acknowledged, how is the service to re-create a public image of security from such penetrations? This question must be answered for very practical operational reasons. Even if the service is entirely certain that it has eliminated the problem, its public image of having been penetrated persists. A potential recruit in a target country, especially if he is well placed and informed, is likely to know about the penetration. Can he afford to trust a case officer from that service to recruit him? How can he be sure that, although one mole has been detected, others do not remain? Unless he can be sure that there are none, he would be unwise to allow himself to be recruited.

Both the Ames case and the Robert Hanssen case of a KGB recruitment within the upper levels of the CIA and the FBI have created precisely this problem for both organizations as they seek to recruit high-quality agents. High-level penetrations of the NSA and NIMA cause terrible damage, but they do not have the same effect on those agencies as they do on clandestine human intelligence and counterintelligence operations because neither NSA nor NIMA recruits agents.

Considering Solutions

The second problem I have outlined—the need for recognition of genuine accomplishments and effective service by clandestine service personnel—is a management problem, not one that submits to structural solutions. It is somewhat less a problem in military human intelligence organizations because the personnel are mostly commissioned officers. Their rank, at least, is an outward sign of achievement. When they retire, they can be honored with formal military ceremonies. Civilian human intelligence services face a more difficult challenge in providing recognition.

The first problem, arising from the syndrome of clandestine operational skills affecting the flow of accurate and complete information up and down the management lines, has no easy solution. It, too, is a management problem, but a very special one. Two ethics are in conflict: the ethic of operational dealings with agents and the ethic of management integrity. The nature of the conflict in a clandestine service is certainly special, but variants of it are found in military organizations. Small groups form very tight social relations in military units, and loyalty to these groups frequently conflicts with the military units' missions and values. The most conspicuous examples, but by no means the only ones, are found in the cheating scandals that occasionally erupt in the military academies. Honor systems in the academies have the practical purpose of socializing young men and women to place the institution's interests and values above their personal and small-group loyalties when there is a conflict. Military operations require that. When a military unit receives a combat mission order, that mission becomes more important than the life of any member of the unit. Its commander must be willing to risk lives to achieve it. This, of course, is the extreme case, but unless military organizations can sustain their priority of values, they will fail in their missions. The

whole idea of an officer's honor and the code for an officer's behavior rest on the recognition that he must live and die by that priority
of values. It is highly undemocratic, seemingly unfair, but finally
essential. And years of indoctrination are required to instill the
ethic. Moreover, the ethic inevitably erodes over time and requires
measures to revitalize it.

There may be a leadership lesson here for controlling clandestine human intelligence organizations, in dealing with the so-
called culture problem. There is no intense socialization experience in the training of clandestine service officers of the kind found
in the precommission training of military officers. It is also worth
examining the military clandestine units for their experience in
coping with the traditional problems of integrity in answering to
the chain of command, recognition of achievements, and so forth.
In principle, the commander of military human intelligence organizations has available methods for dealing with these problems
that are unavailable to the Directorate of Operations.

The third problem, tensions between counterintelligence
concerns and case officers' desire to trust their agents is also a management problem with no structural solution. It can be mitigated
but never entirely eliminated.

The fourth problem, recovering from a high-level penetration such as the Ames or Hanssen case, has a solution, but it seems
so radical that senior intelligence officials have never been willing
to use it. The solution is analogous to the one used in dealing with
mad cow disease in Britain and parts of Europe a few years ago. The
penetrated agency can simply be dissolved, put out of business. If it
still has penetrations that have not yet been discovered, this step
will neutralize them. In the place of the abolished agency, a new
one has to be built with no transfer of veterans of the previous organization, or at most a small and highly vetted set of them. The
problems involved with dropping agents or transferring them to

new case officers are immense, not to mention the costs of starting over in a new organization.

Still, this remedy should not be automatically ruled out. It has been applied in at least one case of a fairly large component of the army's clandestine service several years before it was dissolved and turned over to DIA. The process took more than three years for its administrative execution and then several years thereafter to return to the former level of collection success. As painful, difficult, and costly as it is, after the Ames and Hanssen cases, it is difficult to see why serious professional clandestine service officers would not favor such a measure to restore the health of both the DO and the counterintelligence part of the FBI.

Organizational Issues

If the recommendations for structural reform in the management capabilities of the director of central intelligence are implemented, if the National Imagery and Mapping Agency is allowed to operate as an effective national collection agency for imagery intelligence, and if the programs now in the National Reconnaissance Office support the National Security Agency and NIMA under their national-manager directors, the internal composition of the CIA will change dramatically. The CIA now consists of three major components, the Directorates of Intelligence, Science and Technology, and Operations. The DI, of course, would move under the direct control of the National Intelligence Council and the DCI. A large part of the DS&T consists of parts of the NRO. But it also contains elements that provide technical support to the DO, it manages the Foreign Broadcast Information System, and it has a few other activities.

The proposed organizational changes would leave the CIA with the Directorate of Operations as its primary and major component, but it would also retain a truncated Directorate of Science and Technology. The implications of these changes are clear: the CIA would become a human intelligence agency. Leaving the Foreign Broadcast Information System in the CIA makes sense; it is mainly an overt human intelligence collection effort. Its value extends far beyond the intelligence it provides. The American university and think-tank community is highly dependent on it for access to the media in countries throughout the world. The private-sector research based on the system's products is enormously valuable to the Intelligence Community.

Recommendations

• Restructure the CIA, giving it two major components, the national clandestine service and a component for handling overt human intelligence.

The director of this restructured CIA becomes the national human intelligence manager. His core responsibility is clandestine human intelligence and covert action, but he also must take responsibility for overt human intelligence. To the degree that he is aware of how much can be done through overt sources, he will be better able to target clandestine collection of those requirements that overt sources cannot meet. This kind of trade-off in allocating collection resources is analogous to the kinds of trade-offs the national managers of signals intelligence and imagery intelligence would be asked to make.

• Retain a residual science and technology capability in the clandestine service for support to human intelligence.

The Directorate of Operations requires considerable science and technology support for clandestine operations. Some kinds of science and technology support for CIA operations can and have been provided by the National Security Agency. Taking advantage of that source and others in the Intelligence Community should be standard practice. But the capability to provide primary support must be maintained at the CIA.

At the same time, the CIA needs some science and technology capabilities for a number of collection activities that depend heavily on human intelligence but also have a significant technical component. Some are interagency in character, and maintaining that kind of interagency cooperative effort is absolutely essential.

A frequent defense of the National Reconnaissance Office and the current CIA Directorate of Science and Technology is that they have proven innovative, flexible, and able to field new technical systems rapidly, while similar endeavors within the military services have foundered on bureaucratic infighting and program delays. A few decades ago, this was a sound case. The air force probably would never have produced the U-2 or the SR-71, certainly not as rapidly. A lot has changed over the years, however. The concentration of strong technical skills in both the NRO and the CIA has declined. The National Security Agency's record of rapid development in a number of areas has been impressive, and an effective National Imagery and Mapping Agency could provide the innovation and momentum for fielding advanced imagery intelligence collection platforms.

Finally, the kind of science and technology section prescribed in the Community Management Staff for the DCI would provide a mechanism for virtually unconstrained innovation in temporary research and development programs contracted to a private firm for a limited number of years to prove or disprove the viability of new technologies. It should be temporary, its funding limited to a fixed number of years so that it does not take on a permanent life of its own. The approach could also exploit the residual science and technology capability in the CIA. In other words, with the changes recommended for research and development management by the

DCI, there would be mechanisms for a renewed R&D vitality in the Intelligence Community that has abated over the years in the NRO, the NSA, and elsewhere in the Intelligence Community.

• Formally give the National Clandestine Services of the Directorate of Operations operational control over military clandestine human intelligence operations.

This means making the DO truly take charge of military clandestine human intelligence, treating it as an adjunct to the DO's own efforts. The change actually requires no new formal authorities for the DO. Its present coordination and approval authority over military clandestine operations gives it the appropriate level of involvement.

Certain tensions could result from this change, but they need not. If the Directorate of Operations attempted to allocate military capabilities wholly to nonmilitary collection requirements, ignoring military needs, the Joint Chiefs would object. The DO has long accepted, somewhat ambiguously, a commitment to meet military requirements, but in practice it has never devoted serious resources to them. During the Cold War, for example, U.S. forces in Europe needed a plethora of low-level agent support in then-communist Eastern Europe. The CIA never had the resources to begin to meet them, and that was one reason for retaining army clandestine human intelligence. With operational control over those capabilities, the DO would be much better prepared to deal with such requirements. Likewise, in the Persian Gulf War, and especially in Afghanistan, similar requirements existed but the CIA was without assets to meet them. Had the CIA taken operational control of the army's intelligence support activity and other capabilities in the early 1980s, it could have created those assets without harm to its nonmilitary obligations. Instead, it engaged in a whisper campaign in the congressional intelligence committees and elsewhere to discredit this small unit because it was created to collect intelligence in

Iran in preparation for the hostage rescue mission in 1980 when it became clear that the CIA could not or would not provide it.

A strong argument can be made for the abolition of all military clandestine capabilities. For that argument finally to be persuasive, however, the DO would have to be placed under the operational control of the secretary of defense and of the Joint Chiefs. Otherwise, the experience of the military during the Iran rescue mission would be the norm. No human intelligence support would be forthcoming, and the military could do little about it. The alternative, therefore, is for the DO to take seriously its authorities and potential for managing military human intelligence.

• Allow the CIA/DO to retain its status as the covert action agency, but make it dependent on the Defense Department's capabilities for the conduct of any paramilitary covert actions.

For a time in the 1990s, the need for covert action in general and paramilitary forces in particular appeared to have declined. It reasserted itself after a few years, the case of support to anti–Saddam Hussein Kurds serving as an example. Though less covert, paramilitary operations surged in Afghanistan in the war against the Taliban and al Qaeda. The clandestine human intelligence service is by far the best-positioned to manage such operations in peacetime, heavily dependent as they are on political and military intelligence. At the same time, the DO is unlikely to match the depth of skills and capabilities for paramilitary operations that can be maintained by the military services. Those capabilities now have their own unified command, U.S. Special Operations Command. In peacetime, there is no practical reason why task-tailored complements of special operations forces cannot be passed to the DO's direct control on a case-by-case basis for actual employment. It has been done on a partial basis before. It makes good sense, therefore, to shift the responsibility for creating paramilitary forces entirely to the military services while leaving their peacetime operational employment to the DO. The objections by the secretary of defense on political grounds are understandable, but they should not be the deciding factor. The record includes too many amateurish CIA paramilitary operations. As I have already noted, however, the experience with Special Forces in Afghanistan seems to have changed the attitude at the top of the Pentagon.

• Take a broad approach to designing and implementing CIA management of overt human intelligence.

It is easy to suggest that the CIA, as the national human intelligence agency, take charge of the highly fragmented and poorly managed exploitation of overt human intelligence. It is another thing to be specific about how that should be done. The first problem is finding a practical definition of what is to be included as overt human intelligence. The definition could be expansive, involving all the open media. Clearly the CIA cannot take charge of reading the newspapers for policymakers and military commanders; nor can it watch CNN's battlefield reporting for them. Prisoner of war interrogation units in the army could conceivably fall under CIA operational control, along with many Defense Department debriefing programs. Embassy diplomatic reporting can hardly be the management responsibility of the CIA. Military attaché reporting might well be. These examples suffice to make the point: overt human intelligence collection is a burgeoning area whose management has long been neglected. It needs a disciplined management review that draws practical lines around what is to be included, what is to be excluded. And its leaders will need to work with the NIC and the DI in order to distinguish between what is collection and what is already fit to be treated as finished analysis. For example, some academic works and contract studies are finished analysis, while open-source publications that do not provide finished products for answering specific requirements would be items for collection.

As mentioned in the first recommendation above, an important result of better management control of overt human intelligence collection is that it can show the DO what not to collect through clandestine means. At present, the Intelligence Community has no system for even considering how to determine this, much less accomplishing it.

• Address the CIA Directorate of Operation culture and related problems with a wide range of management, leadership, and organizational reforms, including consideration of disbanding the DO and creating an entirely new clandestine service.

This is, of course, a management recommendation, not a structural one. It is included here to identify a cluster of problems, problems that are poorly understood, even in the Intelligence Community. Although how they can be dealt with effectively is beyond the scope of this book, diagnosis of them should stimulate useful thinking about leadership and management solutions.

Some aspects of this review will meet serious objections from the CIA. Others will be dismissed as nonessential. One or two may be welcomed. The negative reactions should be met with skepticism yet understanding. If the whole set of recommendations were implemented, the CIA, as it has existed to date, would become a thing of the past. It would lose its grip on the DCI. It would lose its so-called leverage on the signals intelligence and imagery intelligence budgets with the dissolution of the National Reconnaissance Office. Finally, it would no longer have the Directorate of Intelligence. These changes might be initially viewed as amounting to the abolition of the CIA.

On the other hand, the clandestine service stands to gain a great deal. Its operational control of military clandestine capabilities could double its resources and greatly extend its reach. It would be much better positioned to have cooperative relations with the military services. Its access to Defense Department paramilitary capabilities would vastly improve its operational capabilities while allowing it to avoid the recruiting, training, and equipping of such forces. The potential for exploiting overt human intelligence collection would be improved and possibly open a number of new programs and operations.

The recommendation that the "culture" problems in the Directorate of Operations be subjected to some far-reaching management responses would be met with stubborn resistance. For the serious long-term interest of the clandestine service, however, that reaction would be a profound mistake. The DO has many accomplishments in its record, and its present and former members are justly proud of them. They compose as gifted and talented a group of people as there is in the government. The sacrifices and risks they have made and the dedication to duty they have shown deserve our highest respect. All that said, however, events have dealt the CIA a series of blows that approach its limit to survive.

The damage began with the congressional investigations in the mid-1970s and has continued with embarrassing disclosures in steady succession ever since, right down to the Ames case. Modest changes are not enough to begin a genuine revitalization of the clandestine service, especially as it continues to pursue what is likely to be a prolonged struggle against al Qaeda and similar organizations which use terrorist tactics against U.S. interests. The changes must be so dramatic, so radical, that they inspire a new sense of confidence based on compelling and demonstrable evidence.

For example, the occasion of major restructuring in other parts of the Intelligence Community provides an excellent opportunity for the CIA to go far beyond moderate internal reform. Can the record of embarrassments ever be entirely washed away from the CIA's public image? Not likely. Why, then, does it not make

sense to take the occasion of Intelligence Community–wide reform and formally drop the CIA label? It could be abolished to
symbolize publicly the depth of change the clandestine service is
undertaking. A new incarnation, marked by a new name, has much
to be said for it—if indeed it is accompanied by the kinds of fundamental reforms recommended here. A professionally serious assessment of the situation inexorably pushes one toward such a
remedy.

8

Spying on Spies
Counterintelligence

ounterintelligence is the most arcane and organizationally fragmented, the least doctrinally clarified, and legally, and thus politically, the most sensitive intelligence activity. I have already made several recommendations for reforming counterintelligence in the Defense Department, and I will not repeat them here, but I will fit them into the proposals in this chapter. Reforming counterintelligence within the overall Intelligence Community, a few fundamental structural reforms have long been imperative although politically unfeasible. Over the past couple of years the political climate has radically changed, making these reforms at least conceivable.

Some historical perspective on the poor performance of U.S. counterintelligence will reveal that repairs need to be made, and not only minor ones. Then a review of the relevant doctrinal assumptions will follow, emphasizing those that can help identify the key problems and guide our thinking about solutions.

The Miserable Record of U.S. Counterintelligence

Before the discovery in 2000 that Robert Hanssen, a well-placed official in the Federal Bureau of Investigation (FBI), had been a Soviet (and later Russian) agent since the early 1980s, reforms in that organization were simply unthinkable. Because the FBI has the primary responsibility for counterintelligence within the domestic United States, no comprehensive reforms of counterintelligence in the Intelligence Community were possible either. Through its long-standing public relations campaign, begun by J. Edgar Hoover, the Bureau built up an image of invincibility and legendary performance. Its success in feeding stories to selected journalists to burnish its image is notorious within intelligence circles. So, too, is its influence in the Congress, which has allowed its directors to shake off numerous embarrassing law enforcement disasters, to keep presidents afraid to investigate the FBI, and to refuse to cooperate with the CIA and the Pentagon on many counterintelligence activities. Now all of this may be changing.

The probable damage Hanssen has done, not only to the FBI's operations but also to the CIA's counterintelligence efforts, may rank as the worst single case in the history of espionage against the United States. It made the FBI look positively amateurish against the Soviet KGB. The events of 11 September 2001 were another major blow to the FBI because intelligence against terrorism in the United States is also the FBI's responsibility. These episodes, combined with continuing disclosures of clumsy and incompetent operations during and after 11 September, have dissolved the FBI's invulnerability to serious criticism. Several major newspapers and respected members of the Senate and House of Representatives have shown an uncharacteristic willingness to speak out candidly about the competence of the FBI in both counterintelligence and counterterrorism in ways that were inconceivable a few years earlier.

The counterintelligence record of the CIA has also been seriously marred, as we have seen. The KGB's recruitment of Aldrich Ames, a senior Directorate of Operations case officer, may have done as much damage to the CIA as its recruitment of Hanssen did to the FBI. The lack of intelligence warning before 11 September, of course, hurt the CIA's reputation as much as it did the FBI's.

The FBI and the CIA are not the only organizations with the counterintelligence capability to run operations. Each of the three military departments in the Pentagon—army, navy, and air force—conduct such operations. Their records of finding hostile intelligence penetrations, not only human agents but also technical penetrations, are less than impressive. Over the past two decades, at least a dozen U.S. military personnel have been discovered to be foreign agents. Because few of them have been in high positions with access to the most sensitive information, however, they have not caused damage comparable to that done by Ames and Hanssen. There have been exceptions. John Walker, a naval petty officer, stole cryptologic keying material and sold it to the KGB for many years, allowing the Soviet Union to collect and break the U.S. Navy's fleet operations communications. An army sergeant in Europe, Clyde Lee Conrad, sold the army's war plans for defending West Germany to Hungarian agents who passed them on to the KGB in Moscow. Most others passed less damaging material to hostile intelligence services.

In Europe, the United States was subjected to massive hostile intelligence efforts, especially in Germany, where the East German services controlled scores of operatives and moved about with great freedom. Thus the number of agents who remain undetected is undoubtedly large. In Latin America, Cuban and Panamanian operations achieved success against U.S. Army units in Panama. In the Far East, excluding the period of the Vietnam War, the most hostile penetrations (mainly by Soviet, Chinese, and North Korean operatives) were made in the local populations of Japan, Taiwan, and South Korea, far fewer among U.S. military personnel.

It is, of course, impossible to know the complete record because undetected agents are precisely that—undetected—and thus uncountable. Reaching a reliable assessment of the effectiveness of U.S. counterintelligence is therefore impossible. Still, several decades of counterintelligence efforts yield a discouraging picture of the quantity and quality of hostile operations against U.S. interests. The end of the Cold War has brought several disclosures about Soviet and East European intelligence operations that improve our understanding of that picture but make it look worse, not better.

The accumulating evidence is most damaging to the FBI. It reveals that the widespread impression of the "good old days" of J. Edgar Hoover's FBI, when it unfailingly caught the bad guys, are a myth. Soviet intelligence, using the Communist Party of the USA as a support organization, simply ran over the FBI again and again during the Cold War and earlier. By the time of World War II, scores of Soviet agents were gaining access to the upper reaches of the U.S. government.[1] Tipped off by army code breakers (who produced the Venona files) with the names of more than two hundred likely agents, the FBI proved unable to collect sufficient additional evidence to indict and prosecute them.[2] Julius and Ethel Rosenberg, who passed secrets about the U.S. atomic bomb program to Soviet officials, were the only exceptions. In 1951, when Whitaker Chambers and Elizabeth Bentley made their disclosures before Congress that Alger Hiss, Harry Dexter White, and many other government officials were passing state secrets to the Soviet Union, the American Communist Party was able to stimulate a fairly successful defensive campaign in the U.S. media. With the new documentation and testimony available from Russia, the guilt of Hiss, White, and others is no longer in doubt. Nor is it possible to dismiss out of hand the wider Soviet penetration of the U.S. atomic energy program, going beyond the Rosenbergs to include illustrious scientists like Robert Oppenheimer and others.[3]

Effective FBI counterintelligence might have secured indictments of scores of Soviet agents during the 1940s and 1950s. Instead, the FBI contributed to the Senate hearings conducted by Joseph McCarthy, whose unconscionable tactics were exploited as a screen for the activities of well-placed Soviet operatives and the American Communist Party. Counterintelligence did not improve in the following decades, not just in the FBI but in the Intelligence Community as a whole. In the 1970s William Kampiles was selling information about U.S. satellite reconnaissance capabilities to the KGB. At the time these so-called "national technical means" were key to verifying arms control agreements between Moscow and Washington. Clyde Lee Conrad was selling the army's war plans for Europe about the same time and would continue to pass secret materials into the mid-1980s. John Walker and his son were selling the navy's communications codes during this same period.

In the 1980s, Ronald Pelton, after he resigned his job at the National Security Agency, contacted the Soviet embassy in Washington undetected by FBI surveillance, and met KGB agents in Europe. There they used James Bamford's book about NSA, *The Puzzle Palace*, to test his bona fides and guide his debriefings. Although Pelton was eventually discovered and successfully prosecuted, the FBI, against pleading by NSA personnel, gave him an easy chance to avoid an indictment. Impatient at surveilling Pelton until he could be trapped in the act of working for the KGB—evidence essential to a successful prosecution—the FBI decided to confront him directly. Had he exercised his constitutional right to refuse to talk, he would be free today. Instead, he believed he could persuade the FBI to pay him to act as a double agent against the KGB and make money both ways. Thus he voluntarily provided the evidence for his own conviction! Similar bumbling amateurism produced a different result in the FBI's handling of the Wen Ho Lee case at the Los Alamos National Laboratory in the late 1990s. Whether Lee gave secrets to Chinese officials may never be known, but the FBI's handling of his case made it easy to sympathize with Lee.

Many other such cases have occurred. Most of them have never been publicly reported, but those that have are sufficient to make the basic point: the FBI has a disastrous record of finding and convicting foreign agents. Moreover, most of the arrests of lesser-known agents have been based on the work of military counterintelligence officers who uncovered them abroad and tracked them for years, patiently building the evidence for a conviction, then enticing them to travel to the United States, where only the FBI has the authority to make the arrest and support the prosecution. These military officers received not even a thank-you as the FBI held press conferences to claim the cases as its own work.

The CIA's counterintelligence record is also mixed, going back to the 1950s and 1960s, when James J. Angleton was its chief. From the paralyzing paranoia of Angleton's reign to the neglectful times in which the KGB defector Yurchenko redefected right under the nose of his handlers, and Aldrich Ames betrayed more than a half-dozen CIA agents working in the Soviet Union, one kind of poor performance replaced another. The counterintelligence record of the military services is also mixed. A number of cases have been handled well, but others demonstrated wholesale incompetence.[4]

It is easy to condemn the FBI and other counterintelligence organizations for not detecting hostile intelligence penetrations early. Moreover, inevitably a few Americans will always be willing to commit espionage against the United States. Is it possible, therefore, that the selective presentation of the record above indicates no more than business as usual, a level of damage that is to be expected? Judgments differ on this matter, but as it will become clear in the following diagnosis, constructing a much improved counterintelligence capability is possible.

Doctrinal Assumptions and Their Ambiguities

In the chapter on principles and concepts, counterintelligence is defined as intelligence gathered about an adversary's intelligence activities and capabilities. In other words, it is no more than collecting information to unmask adversarial intelligence operations and capabilities. This is an important definitional boundary because counterintelligence is often understood to include far more, specifically the security function, involving specific types of action to prevent or neutralize hostile intelligence successes against U.S. interests. Yet as I have made clear, security is a command function in the military and a policy-management function in civilian agencies. Counterintelligence provides information on which military commanders and civilian agency managers should base their decisions regarding security measures, but counterintelligence officials do not have the administrative or command authority to promulgate security policies or to direct security measures. They do, of course, have that responsibility within intelligence organizations, and it includes establishing rules for access to intelligence information by individuals in other agencies. In other words, authority to grant security clearances for access to specific intelligence belongs to the director of central intelligence even for individuals outside the Intelligence Community in user organizations. Beyond this authority, which is really to protect intelligence, not to provide general security to outside organizations, Intelligence Community officials have only the power of persuasion about security measures in nonintelligence organizations.

This analytical distinction, setting counterintelligence apart as purely intelligence and not a security measure, has other practical applications. Deception operations are an example. Valid and comprehensive counterintelligence is imperative for operations intended to mislead an adversary. Because counterintelligence organizations obviously are in the best position to carry out some aspects of deception operations, they become involved in operations that exceed the narrow definition of counterintelligence. Deception operations, like security measures, are command and management functions, not counterintelligence or intelligence functions. Just as tactical intelligence supports military combat operations, counterintelligence must support deception operations. Another example is neutralization of a hostile agent by arresting him or exposing his activities. That is not an intelligence operation but a defensive or security action. Yet in practice it is generally left to the decision of counterintelligence officials, though policy or command approval is always appropriate and often required.

Another doctrinal boundary lies between counterintelligence and "arrest authority." Arrest of a hostile agent certainly amounts to neutralization, but having and using the authority to arrest is not the same as making the decision to do so. Once an espionage agent is detected, by definition a crime is also detected. Law enforcement officials must arrest that person if it is decided to neutralize his activities. Within the military counterintelligence organizations, army counterintelligence does not have arrest authority, but the Naval Investigative Service and the air force's Office of Special Investigations do. Military arrest authority, of course, applies only to military personnel and installations in the United States. Beyond U.S. territorial borders in a war zone, this authority is more extensive for the military services, but in the United States the military services cannot arrest hostile intelligence agents, either U.S. citizens or foreigners, outside military installations. Only federal and local law enforcement agencies can do that. Thus military counterintelligence is inexorably entangled with the FBI and sometimes with other civilian law enforcement agencies.

The key point about this boundary is that counterintelligence organizations need not have arrest authority. In fact, giving it to them unnecessarily expands their responsibilities beyond intelligence work, pushing them precisely where they do not belong: into law enforcement and security work. If we wanted to adhere to the doctrinal principle, then, we would keep counterintelligence organizations separate from law enforcement responsibilities, including arrest authority. At the same time, these organizations must be aware of such law enforcement concerns as the admissibility in court of evidence of acts of espionage. If an arrest is made before adequate admissible evidence has been collected, the subsequent prosecution will fail. Because the counterintelligence organization will most likely be the only group aware of the espionage operation, it will have to decide when an arrest is appropriate. That cannot be left to law enforcement agencies.

Failure to keep this criterion in mind has allowed more than a few Americans working for hostile intelligence services to remain free after they have been discovered. Felix Bloch, a foreign service officer caught in Vienna delivering documents to a KGB agent, escaped trial for such reasons. Ronald Pelton, had he merely refused to talk, would be free. With effective legal counsel available to counterintelligence organizations, such cases should be rare. The army's operation that trapped Clyde Lee Conrad in Germany is an excellent example of how this criterion was met for German courts, where prosecutors and judges were notorious for failing to convict indicted agents.[5]

One final point about doctrine deserves attention. Counterintelligence is widely thought of as human intelligence work. Double agents, spies, and traditional espionage tradecraft are the images evoked by counterintelligence not just in the public mind but also in the FBI and other parts of the Intelligence Community. To be effective, counterintelligence must also involve taking advantage of signals and imagery intelligence to discover hostile agents. And it must discover more than just agents. Counterintelligence operations must also learn about the capabilities and targeting of hostile signals and imagery intelligence. This broader approach is sometimes called "multidisciplinary counterintelligence." Not only the FBI but the CIA and some of the military counterintelligence organizations fail to take a multidisciplinary approach. The personnel in these organizations have been very slow to adapt to the changing technologies of the past three or four decades.

At the same time, counterintelligence as a separate discipline is close to human intelligence, a fact that entangles human intelligence operations with it in a messy fashion. The best window into a hostile intelligence service is a well-placed agent within it. That is why Ames and Hanssen were so valuable to the KGB. A counterintelligence service, therefore, inevitably engages in the same kind of collection techniques as does any clandestine human intelligence service. It recruits agents in the target country, but specifically inside its intelligence services. When counterintelligence is mixed with other clandestine operations, as it is in the CIA, issues of priority arise. Typically, counterintelligence gets second priority when there is a conflict. In some operations, the two efforts can be complementary rather than conflicting. The point remains, however, that a clear boundary between clandestine human intelligence and counterintelligence is difficult to maintain in practice, a fact that must be kept in mind in designing structural improvements in U.S. counterintelligence.

Problems and Solutions

Although many problems in counterintelligence need attention, only two major ones require structural solutions. The first concerns how organizational responsibilities adversely affect professional skills and institutional culture. The second concerns fragmentation that leaves large gaps in the Intelligence Community's overall counterintelligence coverage. From the brief and admittedly eclectic foregoing assessment of the counterintelligence performance record, both problems should already be apparent.

The first is manifest in the FBI and the counterintelligence organizations of the air force and navy—the Office of Special Investigations (OSI) and the Naval Investigative Service (NIS), respectively. All three mix counterintelligence with law enforcement.

Law enforcement techniques that work against criminals seldom work against spies. With a few exceptions, criminals do not enjoy large institutional support systems. They do not have entire governments behind them. Criminals tend to be in a hurry, getting rich or committing murders or both, and then moving on. Except in corporate and white-collar crime, criminals tend to be neither well educated nor highly intelligent. The FBI predominately uses three methods against criminals: telephone taps, informers in criminal circles, and heavy-handed interrogations.

Although these tactics often work against criminals, they do not work against spies and terrorist organizations. Spies and terrorists have motivations and goals that differ dramatically from those of criminals. Spies are normally well financed by their governments; terrorists often have wealthy nonstate organizations behind them. Foreign intelligence officers tend to be highly educated and trained for years in espionage techniques. So too are most managers of terrorist operations. They know U.S. laws and the rights they afford arrested suspects. Foreign intelligence officers can retreat to the secure space provided by countries' embassies. When their spies flee the United States, foreign governments will rarely honor the United States' extradition requests.

Law enforcement agencies thrive on media coverage. Counterespionage agencies fear it because it can destroy or neutralize their operations. Law enforcement agencies are eager to make an arrest. Counterintelligence operatives must be patient while tracking spies until they create enough evidence to support a conviction. This may require a year or two, sometimes longer. Counterintelligence officers must know foreign languages, cultures, and ideologies. Rare is the police agent who does.

Neither the FBI nor the OSI and NIS can maintain both criminal-catching and spy-catching organizational cultures. The two are simply incompatible. The crime-fighting part of all three organizations is dominant; competence in counterintelligence is secondary. More than a few veterans of the FBI's counterintelligence division have complained about the phenomenon in their organization. The records of the OSI and NIS reveal essentially the same pattern. Both organizations have botched high-profile cases. The NIS, for example, was alerted in the mid-1980s to the case of a marine guard in the American embassy in Moscow. The guard reportedly was cooperating with the KGB, allowing its operatives access to parts of the embassy building at night. Although most evidence suggested that the marine was guilty, the NIS's interrogation techniques and other actions soon destroyed the case for the prosecution. The handling of the case made much of the evidence inadmissible in court and undercut the NIS's credibility with the jury.

The CIA and the army's counterintelligence service have no law enforcement responsibilities. The army also separates counterintelligence entirely from other clandestine human intelligence, but at the CIA, the Directorate of Operations does not keep such a clear boundary. Both organizations, of course, have mixed records against penetrations, but the example of the army's operation to discover Clyde Lee Conrad is especially impressive, and the CIA counterintelligence division gave the army important assistance in some aspects of the case. The vastly greater resources committed to this operation, the level of skills, and the degree of patience shown by those officers who ran it distinguish it sharply from the way the FBI typically manages such cases.

The second major problem, fragmentation of counterintelligence coverage, can be explained more briefly. Although many government agencies need counterintelligence support in assessing and managing their own security, only five—the FBI, the CIA, and the three military departments in the Pentagon—actually conduct counterintelligence operations. These are five separate and largely disconnected worlds, and none is in a position, or willing, to integrate them and provide a comprehensive national counterintelligence view of all hostile intelligence (and terrorist) operations against the United States.

Counterintelligence relations between the FBI and the CIA have been conflict-ridden from the beginning. When William Donovan created the Office of Strategic Services during World War II, the strife began, even though the OSS, taking the lead in Europe and East Asia, left Latin America to the FBI, which had first put agents abroad there. A number of memoirs and histories catalogue the bitter competition between the CIA and the FBI in the CIA's first decade.[6] The CIA, of course, has counterintelligence responsibility abroad, and the FBI has it within the United States. The idea of combining their separate pictures of hostile intelligence threats is naturally unacceptable to both, although in recent years centers set up by the DCI for intelligence about terrorists and narcotics trafficking have included both the FBI and the CIA. What each exposes in these centers, of course, is far short of the comprehensive picture from their holdings. The FBI has used its counterterrorism and narcotics-trafficking tasks to justify placing more of its agents abroad in U.S. embassies. Thus the fragmentation and competition are no longer limited to the United States but extend into new offshore areas.

The contentious relations between the FBI and the CIA, combined with the autonomy of the military department's counterintelligence operations, leaves large gaps in coverage. The DCI cannot produce a comprehensive operational picture of either the hostile intelligence threats or the terrorist threats to the United States. None of the five agencies has a complete operational picture. Intelligence Community "centers" involving all agencies can never do more than paper over the basic structural problem. This problem, which has existed for decades, should have been painfully apparent after Robert Hanssen was unearthed as a Soviet agent in the very heart of the FBI, but no serious voices in Congress or elsewhere pointed it out and called for facing up to it. Only after 11 September 2001 did it become impossible to hide or ignore. Still, though some members of Congress and newspaper editorial writers have expressed a new awareness of the problems, no serious steps toward closing the gaps have yet been proposed.

The solution to both of these problems—1) organizational incompatibilities between law enforcement and counterintelligence operations, and 2) fragmentation in counterintelligence coverage—is the creation of a national counterintelligence service. For purposes of brevity, let us call it by the acronym NCIS.

The first step in the creation of such a service is to eliminate entirely the FBI's responsibility for counterintelligence. Its counterintelligence division would yield its files, records, and agent operations to the new NCIS. A number of the FBI's counterintelligence specialists could well be transferred to the NCIS, but not merely as an administrative action, but rather based on the quality of their skills and past performances.

Second, the NCIS must be made responsible for coordinating all of the counterintelligence operations of the CIA, army counterintelligence, NIS, and OSI. These four agencies would continue with their operations just as in the past, but each would now have to open its records and files to NCIS officials, allowing them access to its operational view of hostile intelligence threats. This authority would permit the creation of a comprehensive counterintelligence picture. It would allow the NCIS to close the gaps now extant between the five counterintelligence organizations. The NCIS might eventually take operational control over them all, the way NSA asserts control over the military service cryptologic elements. It will, in any case, need to direct changes in operations to prevent overlap, to improve coverage, and to take other actions to improve the combined effort. This might, for example, include temporarily reallocating agents to support costly surveillance and technical coverage of high priority cases. Material for such training would include a complete set of the detailed records of all hostile intelligence cases dealt with in the past. Strangely, this kind of historical record seems to have little value in U.S. counterintelligence circles.

With coordination authority over all other counterintelligence agencies, the director of the NCIS is positioned to become the national manager for counterintelligence. That makes him responsible for a consolidated national counterintelligence program budget under the DCI's program authority. Today no one is positioned to put together the aggregate counterintelligence resource requirements for the entire Intelligence Community. Traditionally, counterintelligence has been grossly understaffed and short of fiscal resources. That goes some way in explaining sloppy and impatient practices by the FBI and also in the military services.

Third, the NCIS should be placed under the DCI for program management, collection tasking, and Intelligence Community policy. In other words, it should have a status similar to the CIA once the DCI has given up his CIA hat and truly become the leader of the Intelligence Community.

Fourth, create a special court—or, alternatively, use the Foreign Intelligence Surveillance Act court—with authority to grant surveillance authority to the NCIS. It should also have authority to investigate and review NCIS operations that become politically or legally sensitive. To be sure, the intelligence committees of the Congress must also have oversight of the NCIS, but the judicial branch needs to be able to keep an eye on it as well. Questions of the civil rights implications of its operations will inevitably arise in the public's mind, the way the public has periodically expressed concern about the propriety, even legality, of the FBI's operations. Because there will be temptations at times for the NCIS to cut legal corners or to yield to political pressures, it must be monitored by outside authorities.

Fifth, the NCIS should not have arrest authority. The FBI can still perform that service except on military installations and in U.S. military force deployments abroad. Military police can provide that service in those locations. The NCIS will, of course, have to maintain close liaison with the Department of Justice for several reasons, but especially to support prosecutions of those accused of espionage. The FBI will no longer be in a position to do that.

These five points provide the broad outlines of the required structural reforms. Once implemented, they can eliminate the weaknesses caused by mixing counterintelligence and law enforcement in one organization. And they make it possible to close the gaps in counterintelligence coverage, providing a comprehensive national counterintelligence picture.

These changes also should mitigate some lesser problems that have arisen from confusion about the distinction between responsibilities for security and for counterintelligence. Still, they will not clean up all the boundary problems. Several of those simply have to be managed because they cannot be eliminated. Nor do these changes address all the challenges involved in having the NCIS coordinate the military departments' counterintelligence operations and provide at least some support to tactical military commanders.

Recommendations

The following recommendations merely make more specific the foregoing points:

- Create a National Counterintelligence Service (NCIS). The FBI's counterintelligence department can form the core of this organization, and it can be augmented with small elements from the CIA's counterintelligence organization.
- Designate the director of the NCIS the national manager for counterintelligence, responsible to the DCI in the same way as is the national manager for human intelligence, both for operations and for resource management—that is, the national counterintelligence program budget.
- Give NCIS coordinating authority over all counterintelligence operations within Intelligence Community components. Whether this authority should include operational control over all, or part, of counterintelligence operations in other Intelligence Community components should be determined by experimentation and practical experience.
- Give the national manager for counterintelligence responsibility for providing support to all departments and agencies at the national level. To the extent that he assumes operational control of military counterintelligence, he must also provide counterintelligence support to tactical military units. Just as the national managers for signals intelligence, human intelligence, and imagery intelligence are responsible for linking national collection assets to support for tactical military operations, the national manager for counterintelligence can probably usefully link certain kinds of support to augment and strengthen internal capabilities in tactical military organizations.
- Make the national manager for counterintelligence re-

sponsible to the DCI for maintaining a comprehensive
picture of all relevant counterintelligence targeting of
foreign intelligence services.
• Direct the national manager to create a counterintelligence school and ensure that it has available the record
of all counterintelligence cases as its primary instruc-
tional material.

These recommendations are not only compatible with other
recommendations in this book but are necessary to the implementation of several of them. Of all the recommendations, this is perhaps the most compelling and urgent in light of the intelligence
failure to warn of the al Qaeda attacks against the United States in
September 2001. Progress in improving homeland security and especially border controls depends heavily on a far more effective
counterintelligence and counterterrorism capability in the Intelligence Community. Finally, while structural reforms make improvements possible, they do not make them certain. That requires
good leadership as well.

9

Conclusion
What It All Means

I have in the foregoing analysis been highly critical of the Intelligence Community, but one should not conclude that it has been an overall failure or that there is little positive to say about its performance. In a few important instances, it has had embarrassing failures that should not be tolerated. On balance, however, it has performed impressively, at times accomplishing remarkable feats. Most of these successes must remain secret if there is to be a chance of repeating them in the future. There is, nonetheless, sufficient unclassified evidence of the achievements of the Intelligence Community to make the general point.

For example, a case can be made that the Intelligence Community was no less critical for deterring Soviet aggression during the Cold War than were U.S. military forces. In a few instances the Intelligence Community denied Moscow the surprise that Soviet leaders believed was imperative for success, forcing them to call off an operation. That was probably a key reason that Soviet military forces did not intervene in Poland in late fall 1980. The Intelligence Community also verified arms control treaties reasonably well. It provided a surprisingly accurate record of the size and organization of Soviet military forces, weapons characteristics, quantities produced, and peacetime force dispositions. In a number of small U.S. military operations, intelligence support was so effective that the military commanders involved were astounded, although in other instances it was painfully inadequate.

In military history, intelligence chiefs have seldom been praised but have often been blamed for their commanders' operational failures. Even General Schwarzkopf, although his Central Command enjoyed as large an intelligence advantage during the Persian Gulf War as any in history, was quite critical of some of the intelligence support he received. My diagnosis heretofore needs to be seen in the context of what's at stake. No other intelligence system in the world comes close to matching the overall capabilities of the U.S. Intelligence Community, but several countries and groups, such as al Qaeda, have obviously learned to evade some of its collection and detection methods. The challenge for U.S. intelligence today, therefore, is not to avoid being outclassed (counterintelligence aside); rather, it is to deal with several accumulating dysfunctions and inefficiencies that afflict many large high-technology organizations. In particular, it confronts three major problems: a) ineffective management of a constant infusion of new technology, b) changing targets and intelligence requirements, and c) long-standing organizational legacies that obstruct desperately needed change.

The need for reform should also be put in another context. Intelligence performance simply cannot be separated from foreign policymaking and military operations. What initially appears to be an intelligence failure often turns out to have been a failure of interaction between political and military leaders on the one hand and intelligence officials on the other. Too often the users of intelligence simply refuse to recognize the implications of what intelligence officials are telling them, either because bad news is never welcome or because they cannot integrate into their view of reality facts that do not fit their preconceptions.

These perspectives, however, should not be invoked to explain away intelligence failures and to avoid long-needed reforms. Genuine intelligence failures do occur. Preconceptions can prevent intelligence collectors and analysts alike from seeing accurately what is happening. Collectors also can fail to realize that techniques that worked in the past have been made ineffective by opponents' security measures.

The weakness of U.S. counterintelligence is difficult to exaggerate. Signals, imagery, and human intelligence are not without serious flaws. To pretend that lack of warning of the al Qaeda attacks of 11 September was not an intelligence failure is self-deluding. Yet to see it as an intelligence failure is not to absolve political leaders—who are, after all, in charge of the Intelligence Community—of the ultimate responsibility. This disaster is a failure of both intelligence and policy. It should convince professional intelligence officers and top policymakers that reforms can no longer be responsibly dodged or put off, as they were by the Senate's authorization of a presidential commission in the mid-1990s that whitewashed the problems precisely as its sponsors intended.

Many senior intelligence officials, some currently serving, others retired, will reject my reform proposals out of hand. Senior executive branch officials and members of the congressional intelligence committees will also view them skeptically. If they take some time and reflect on the overall results that these reforms could bring, however, some will change their minds. What at first appears to be dismemberment of their favorite organizations, essentially the CIA, the NRO, and part of the FBI, turns out to enhance the professional prospects of the incumbents in those organizations. Many positions would be cut, but probably as many new ones would be created. In counterintelligence, a new National Counterintelligence Service would, without question, require a large increase in personnel over the number now committed to counterintelligence. Moreover, working in the new organizational context should eliminate many of their old frustrations and open up chances for greater accomplishments. Consider the major gains that would accrue to the entire Intelligence Community.

First, the director of central intelligence and all the components of the Intelligence Community would be able to make a far more persuasive case for their budgets. They could present to the intelligence committees in Congress more compelling explanations of how added dollars and personnel will actually produce better intelligence. They would also be able to make stronger arguments for certain program reductions and shifts of resources among programs. The present predicament in Intelligence Community resource management has its origins in the bureaucratic fears and aims of the 1940s and 1950s. The CIA's aim was to prevent the Defense Department from undercutting its determination to maintain an intelligence structure apart from the military. That battle has long been decided. Yet the program management system sustains the mentality and gamesmanship of forty years ago.

Second, reforms are essential if the Intelligence Community is going to cope effectively with the continuing changes in technology. One feather in the cap of the early Intelligence Community organizational innovation is its success in high-altitude photography provided by the U-2 aircraft, followed by satellite imagery and satellite collection of signals intelligence. The National Reconnaissance Office was created with very loose spending rules and highly secret organizational arrangements to press forward in all kinds of advanced collection technologies. In the viscous bureaucracy of the Defense Department's research and development programs, these advances might not have been made, or certainly would not have been made until years later. Over several decades, however, the NRO's once agile and technically competent bureaucratic arrangements have atrophied. At the same time, new possibilities in using the most advanced collection technologies in operations began to be thwarted by aging NRO bureaucratic controls. The NRO provides only the most conspicuous example of the problem of changing technology and static organizational arrangements. Many others can be found within the National Security Agency and the military services' intelligence organizations.

Third, the proposed reforms are in fact a continuation, not a reversal, of trends in U.S. intelligence organization dating back to World War II. Within the Defense Department, the creation of the National Security Agency in the early 1950s was strongly opposed by the military services. Without the consolidation, essentially establishing a national manager for signals intelligence, the subsequent research and development programs in the 1950s would have been delayed or entirely blocked, handicapping the U.S. signals intelligence effort. The National Imagery and Mapping Agency, created in the mid-1990s, should have been established two decades earlier, but again, it marks another trend in line with the doctrinal principle of national managers that I have advocated throughout this book.

Ironically, a national manager system of human intelligence has long been possible without any major organizational changes. Attitudes and concerns with turf boundaries that originated between the OSS's clandestine operations and operational military commanders strangely persist today, blocking the emergence of the director of the Directorate of Operations at the CIA as the national manager. The bases for those legacies have vanished long ago. It is time to forget them.

Counterintelligence has also failed to evolve progressively. In fact, it has remained where it was during and after World War II, unable to stem the onslaught of Soviet and other foreign intelligence penetrations. If the Robert Hanssen case was not convincing evidence of the need for radical reform, then the events of 11 September should leave no doubt. The notion of making law enforcement agencies responsible for counterintelligence (and counter-terrorism) is a formula for failure. Not only would a new National Counterintelligence Service make possible great improvements in operational competence but it should also prove more effective in resource management. An NCIS would be able to demonstrate just how poorly funded counterintelligence has been for decades.

The general trend in the DCI's position has been toward a greater role in leading the Intelligence Community. The proposals made here, especially for separating the posts of DCI and director of the CIA and for providing the DCI with much stronger staffs, should empower the DCI, not weaken the position, as CIA officials have so long feared. In reality, double-hatting the DCI with CIA directorship has prevented its logical and ultimate development.

Finally, the challenge of providing intelligence to a new department of homeland security makes these reforms more compelling. The system of national managers supplying intelligence to intelligence analysis units in many places in the Department of Defense, Department of State, and other agencies would be readily applicable to a department of homeland security. The idea of putting the FBI or CIA in the department of homeland security, as some have suggested, makes no sense at all. Both depend heavily on intelligence provided by the National Security Agency and the National Imagery and Mapping Agency. Would they still do so? If so, then the department would not have its own autonomous intelligence system. If not, would the department of homeland security create another NSA and NIMA? If every cabinet department that uses a lot of intelligence had its own CIA and FBI, then the bureaucratic warfare, fiscal waste, and duplication of efforts would be difficult to exaggerate. Moreover, the quality and effectiveness of intelligence in every department would drop dramatically from what the Intelligence Community can provide today, even with its inefficiencies.

Because terrorist attacks have become such a serious threat, intelligence reforms, especially in counterintelligence, need the highest priority. As our analysis has clarified, patching and repairing here and there in the Intelligence Community will not meet the future demands of homeland security or military campaigns like the one in Afghanistan. Only structural reforms can open the way for the kind of improvements that are needed.

If most of the reforms advocated in this book had been implemented a decade ago, would enough of al Qaeda's preparations for the 11 September attacks have been detected to have prevented them? Obviously a definitive answer is impossible, but the probabilities would have been greatly improved. Basic structural changes in counterintelligence clearly offer the strongest "might have beens" in such speculations. A National Counterintelligence Service without law enforcement responsibility would have had much better prospects for putting together the pieces of information that were accumulating from many sources in the United States and Europe. Moreover, it probably would have been much more effective in getting European governments to take the terrorism problem seriously. The FBI's diplomatic skills in this regard leave much to be desired.

The argument is more speculative for the rest of the Intelligence Community but not easy to discount outright. Had the DCI been supported by national managers in charge of all the resources of the collection disciplines, the National Security Agency could have had better foreign language and collection capabilities. The National Imagery and Mapping Agency also might have targeted its capabilities more effectively. The CIA's clandestine service could have had better access to resources, but it would have needed much stronger backing from top political leaders to have had a chance of detecting preparations for 11 September. Most important, the DCI would have been much better informed about the capabilities in all the collection disciplines, and he would have been better able to direct and coordinate the overall collection effort. In other words, he could have been making a more compelling case for resources over the past decade with the proposed resource management system, and he would have had at his disposal flexible analysis capabilities to focus on the terrorist problem. It would not have been diluted in interagency counterterrorism centers in which the FBI and others did not really want to participate.

Even with more effective intelligence, the president and his cabinet officials would have to have been willing to believe the warning of an attack and act on it. Moreover, given the fragmented character of U.S. border controls and law enforcement capabilities, could the plethora of different agencies involved have acted to stop the attacks even if they had been warned? There are reasons to doubt it. As I have emphasized, the "policy" side of the government is as much at fault for the events of 11 September as is the "intelligence" side. This has been true for a long time. A border management agency, putting together most of the small agencies with responsibilities for different aspects of border controls, was proposed by President Carter as one of his several reorganization prospects in 1978–80. It was defeated by the very set of bureaucratic and congressional interests that oppose creating a homeland security department today. Had it been implemented, the odds of preventing the 11 September attacks would have been vastly increased.

Although structural reform is essential for any major improvements, a number of policy changes are also needed. Elaborating them comprehensively would require another book, but one such change was proposed in the chapter on signals intelligence: shielding of sources and methods for intelligence operations. The analogy cited between the damage done in the 1930s before Pearl Harbor and the laxity and loosening of controls by intelligence officials themselves before 11 September should give both the president and the Congress serious pause, even if clear causal links cannot be established in the latter case as clearly as it was in the former.

Of equal or greater importance is ensuring that no changes in Intelligence Community structure and responsibilities infringe on the civil rights of American citizens. Only in the past six decades has the United States had a large intelligence establishment during peacetime. From the end of the Revolutionary War to the start of World War II, intelligence capabilities created during wartime were abolished or allowed to wither during peacetime. This relatively new development of peacetime intelligence collection and analysis has raised concern over abuse of civil rights and presidential action through covert methods beyond both public and congressional oversight.

Given the staggering international responsibilities that rest on the United States after the Cold War, the needs for effective intelligence are all the greater. Turning back to the practice of earlier times by closing down most of the Intelligence Community in peacetime would be irresponsible. But can the American political system accommodate expanded peacetime intelligence operations without endangering citizens' rights? Like policy questions, this issue is beyond treatment in this book, but my answer is yes. Such adaptation does, however, require strong oversight and controls. Counterintelligence obviously presents the greatest dangers, but excellent counterintelligence operations can be conducted with full protection of civil rights. When intelligence requirements appear to threaten those freedoms, the best guide is to remember that no terrorist or spy operation ever caused the demise of a liberal democracy, but acts of parliaments have brought down a few.

Appendix
Intelligence Organizations and the Intelligence Process

This appendix lists and briefly describes the intelligence process and the major organizations, agencies, and budgetary components of the Intelligence Community.

Intelligence is a vital element of our national security. There is an absolute need for it and for a U.S. Intelligence Community to provide it. Without intelligence our government and military forces would be at the mercy of better-informed adversaries and competitors.

Intelligence can be defined as knowledge or foreknowledge of relevant aspects of the world around us that is used by policymakers and military leaders to make decisions.[1] The intelligence cycle begins when these officials request such information. Collection management involves requesting intelligence on certain questions from the Intelligence Community. Collection is the gathering of "raw" intelligence or data, which collectors sometimes put into a form that can be used by production offices. In specialized collection agencies managers orchestrate technical collection management, which employs diverse collection systems operating on the basis of priority and opportunity. This is a highly complicated and specialized activity that varies among collection agencies depending on the technologies involved. Production divisions of the Intelligence Community analyze the raw and processed data and produce reports and studies, which can be defined as "finished" intelligence. Dissemination is the process of providing the finished intelligence to the consumer, or user. The user then (ideally) provides feedback and additional direction (further collection management) to the collectors and producers, who should in turn provide more and better intelligence. This cycle should result in the Intelligence Community providing timely, useful products allowing informed political and military decisions by U.S. government officials and military officers.[2]

Intelligence Management Structure

The Director of Central Intelligence (DCI)

The director of central intelligence is the statutory head of the Intelligence Community. The DCI is charged with three major tasks: managing the Intelligence Community at large, directing the Central Intelligence Agency (CIA), and serving as principal intelligence adviser to the president. His office was defined by the National Security Act of 1947. Since 1976 the DCI has been assisted by a deputy director of central intelligence (DDCI).[3] In 1996 Congress approved adding a second deputy director, for community management, and three assistant directors of central intelligence, one for collection, one for analysis and production, and one for administration.[4] Several management and coordinating bodies also aid the DCI.

The Community Management Staff (CMS)

The DCI in his role as manager of the Intelligence Community is supported by a Community Management Staff (CMS). The CMS in 1992 superseded the Intelligence Community Staff. The CMS develops, evaluates, justifies, and monitors the budget for the Na-

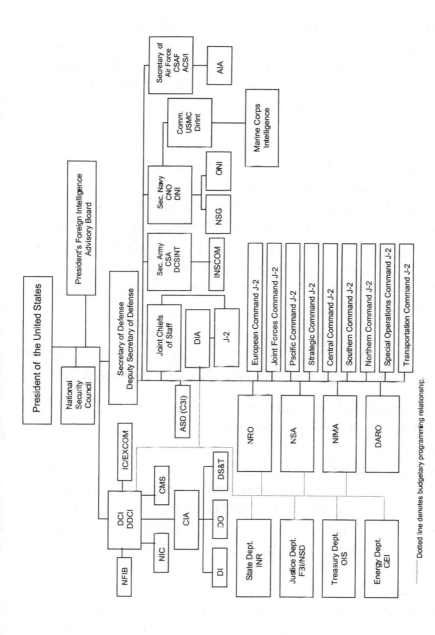

Figure 7. The Current U.S. Intelligence Community

President of the United States

President's Foreign Intelligence Advisory Board

National Security Council

Secretary of Defense
Deputy Secretary of Defense

ASD (C3I)

Joint Chiefs of Staff

DIA

J-2

Sec. Army
CSA
DCSINT

INSCOM

Sec. Navy
CNO
DNI

NSG

ONI

Comm. USMC
DirInt

Marine Corps Intelligence

Secretary of Air Force
CSAF
ACS/I

AIA

European Command J-2
Joint Forces Command J-2
Pacific Command J-2
Strategic Command J-2
Central Command J-2
Southern Command J-2
Northern Command J-2
Special Operations Command J-2
Transportation Command J-2

NRO

NSA

NIMA

DARO

IC/EXCOM

DCI
DDCI

CMS

NFIB

NIC

CIA

DI

DO

DS&T

State Dept. INR

Justice Dept. FBI/NSD

Treasury Dept. OIS

Energy Dept. OEI

Dotted line denotes budgetary programming relationship.

tional Foreign Intelligence Program. The CMS also conducts long-range strategic planning for meeting long-range objectives: translating the needs of intelligence consumers into national intelligence needs; integrating the efforts of the collection disciplines (signals intelligence, imagery intelligence, human intelligence, and measurement and signature intelligence) to satisfy those needs; and evaluating the Intelligence Community's performance in satisfying those needs. The CMS is headed by the executive director for Intelligence Community affairs. Currently there are several main offices, referred to as groups, in the CMS: Resource Management, Program Assessment and Evaluation, Requirements and Plans, Policy and Special Issues, and Advanced Technology. Also resident in the CMS are three Intelligence Community secretariats: the Quality Council, Intelligence Systems, and Intelligence Program Review Group.

The Intelligence Community Executive Committee

The Intelligence Community Executive Committee (IC/EXCOM) in 1992 superseded the National Foreign Intelligence Council, which was established during the Reagan administration. The IC/EXCOM aids the DCI with regard to intelligence policy and resource matters; priorities and objectives for the National Foreign Intelligence Program budget; Intelligence Community policy and planning; and Intelligence Community management and evaluation.[5] Membership on the IC/EXCOM is similar to that of the National Foreign Intelligence Board. The DCI is chairman, and the deputy DCI is vice chairman and CIA representative. Other members include the directors of the National Security Agency, the Defense Intelligence Agency, Intelligence and Research, the National

Reconnaissance Office, and the National Imagery and Mapping Agency; the chairman of the National Intelligence Council; the assistant secretary of defense for command, control, communications, and intelligence; the undersecretary of defense for acquisition and technology; the vice chairman of the Joint Chiefs of Staff; and the executive director for Intelligence Community affairs.[6] IC/EXCOM meetings are not necessarily limited to principals.[7]

The National Intelligence Council

The National Intelligence Council (NIC) serves as a senior advisory group to the DCI and is responsible for determining and promulgating the Intelligence Community's judgments on issues of importance to policymakers.[8] The NIC is the producer of the national intelligence estimates.[9] The NIC includes a chairman; twelve national intelligence officers, each of whom covers a major geographical region or function (for example, strategic programs and nuclear proliferation); and a small staff of supporting analysts.[10] Under the NIC is the National Intelligence Production Board, and several other Intelligence Community–wide intelligence production committees. The NIC was created in 1979; in 1992 it was restructured and moved from its location at the CIA facility in order to underscore its role as an independent, community-wide organization.[11]

The National Foreign Intelligence Board

The National Foreign Intelligence Board (NFIB) is responsible for review and coordination of national foreign intelligence; interagency exchanges of foreign intelligence; arrangements with foreign governments on intelligence matters; and protection of intelligence sources and methods. In practice, most of the board's business has been to review and approve national intelligence estimates (NIEs).[12] NIEs are defined as "the DCI's most authoritative written judgments concerning national security issues. They deal with capabilities, vulnerabilities, and probable courses of action of foreign nations and key developments relevant to the vital interests of the United States."[13]

The NFIB is chaired by the DCI; the board also includes the deputy DCI as vice chairman and CIA representative; the directors of the Defense Intelligence Agency, the National Security Agency, and the National Imagery and Mapping Agency; the assistant secretary of state for intelligence and research; the assistant director for intelligence of the Federal Bureau of Investigation; the special assistant to the secretary of the treasury for national security; and the assistant secretary of energy for defense programs. The director of the National Reconnaissance Office attends as necessary. The senior army, navy, air force, and Marine Corps intelligence officers sit on the board as observers. NFIB meetings tend to be limited to principals only.[14] The NFIB has existed, under various designations, since the creation of the CIA in 1947; it has used its present name publicly since 1978.[15]

Intelligence Organizations

The Intelligence Community has thirteen major organizations that are concerned with gathering raw intelligence, or producing and operating systems that gather raw intelligence, or producing finished intelligence. Many of these agencies perform more than one function.

The Central Intelligence Agency

The Central Intelligence Agency (CIA) was created by the National Security Act of 1947. The act charged the CIA with (among other responsibilities) providing intelligence to the National Security Council, and "perform[ing] other such functions and duties related to intelligence affecting the national security as the National Security Council may from time to time direct." This broadly written directive provides the justification for the CIA's human intelligence collection and covert action activities.[16]

The CIA includes three major divisions: the Directorate of Intelligence (CIA/DI), the Directorate of Operations (CIA/DO), and the Directorate of Science and Technology (CIA/DS&T). CIA/DI is the directorate charged with analysis and production of finished intelligence products, based on "all-source" intelligence collection—that is, from the various intelligence disciplines. CIA/DO is charged with the collection of human intelligence and with conducting covert action. CIA/DS&T operates a number of technical intelligence collection programs, including the Foreign Broadcast Information Service, and provides technical support to CIA/DO. In addition, a CIA Directorate of Administration (CIA/DA) provides administrative support to the Agency.[17]

The Defense Intelligence Agency

The Defense Intelligence Agency (DIA), founded in 1961, is charged with a variety of analytic and collection tasks. Mainly it provides all-source finished intelligence to the Department of Defense for two primary purposes. One purpose is to provide threat assessment in order to formulate requirements for U.S. weapon developers, and to improve training and organization. The other main purpose of finished intelligence from the DIA is support to military operations. The DIA supports the Joint Chiefs of Staff and the unified commands in this capacity with an office headed by a two-star military officer, who serves as director for intelligence for the Joint Staff. The DIA also administers the Defense Attaché System and the Defense Human Intelligence Service, as well as the office that oversees collection of measurement and signature intelligence. The DIA is headed by a three-star military officer, who reports to the secretary of defense.

The National Security Agency

The National Security Agency (NSA) was founded in 1952. A combat support element of the Department of Defense, it is charged with collecting and disseminating signals intelligence, which it provides to military commands, civilian agencies, and producers of all-source finished intelligence.[18] It is led by a three-star military officer, who reports directly to the secretary of defense. The NSA's day-to-day operations fall under its Consolidated Cryptographic Program; the NSA also develops signals intelligence technology. It is the largest intelligence agency in terms of personnel.[19]

The National Imagery and Mapping Agency

The National Imagery and Mapping Agency (NIMA) was created on 1 October 1996. It is a combat support element of the Department of Defense. NIMA collects and provides imagery, imagery intelligence, and geospatial (mapping) information (but not finished products) in support of national security objectives.[20] NIMA is responsible for the functions of the following disestablished organizations: the Central Imagery Office, the Defense Mapping Agency, and the Defense Dissemination Program Office; it also performs the functions of the CIA's National Photographic Interpretation Center. NIMA also has absorbed the imagery exploitation, dissemination, and processing elements of the Defense Intelligence Agency, the National Reconnaissance Office, and the Defense Airborne Reconnaissance Office.[21] By statute, NIMA is led by a three-star military officer.[22] NIMA's director reports directly to the secretary of defense.

The National Reconnaissance Office (NRO)

The National Reconnaissance Office (NRO) was established in 1960. An agency of the Department of Defense, it is charged with developing, acquiring, launching, and operating space-based reconnaissance systems for the entire Intelligence Community. The NRO does not produce intelligence itself; its most direct relationships are with the collection agencies in the Intelligence Community, rather than with intelligence producers or consumers.[23] The director of the NRO is also the assistant secretary of the air force for space.[24]

The Defense Airborne Reconnaissance Office

In 1993 the Defense Airborne Reconnaissance Office (DARO) was formed. DARO is a Department of Defense organization charged with management oversight of the development and acquisition of all joint Military Department and Defense-wide airborne reconnaissance capabilities, including manned and unmanned aerial vehicles, their sensors, data links, data relays, and ground stations. The hardware programs DARO manages are integral components of U.S. national imagery intelligence collection capabilities. DARO is led by a two-star military officer.

Army, Navy, Air Force, and Marine Corps Intelligence

The military services have their own intelligence elements. Army intelligence is headed by the deputy chief of staff of the army for intelligence. He is supported by the U.S. Army Intelligence and Security Command, which combines signals and other intelligence activities, as well as providing cryptologic support. The navy is supported by the Office of Naval Intelligence, which is headed by a flag-rank director of naval intelligence. The Office of Naval Intelligence is responsible to the chief of naval operations for intelligence, cryptology, special security, and foreign counterintelligence. Signals security is handled by the Naval Security Group. Marine Corps intelligence is headed by a director of intelligence, who is the senior intelligence officer and the commandant's principal staff officer and functional manager of all-source intelligence, counterintelligence, and cryptologic matters. Air force intelligence is headed by an assistant chief of staff for intelligence, who manages air force signals, technical, human, and imagery collection efforts.[25] Each service's intelligence is organized somewhat differently. The service intelligence organizations perform a variety of functions, including collection and analysis for force enhancement and support for military operations.[26]

There are nine unified commands that have responsibilities for military operations worldwide: the European Command, Joint Forces Command, Pacific Command, Strategic Command, Central Command, Southern Command, Northern Command, Transportation Command, and Special Operations Command. All have intelligence staff sections.[27] Intelligence staffs or officers are also located at all service organizational levels down to battalion in the ground forces, wing or squadron in the air forces, and individual ships in the navy.

State, Energy, and Treasury Department Intelligence Offices

Besides the Department of Defense, several cabinet agencies have intelligence offices. All are much smaller than the CIA or DIA and serve almost exclusively to provide their agencies with finished intelligence.

The Bureau of Intelligence and Research (INR) produces intelligence for the State Department. It is one of the three intelligence agencies (the others are the CIA and DIA) that produce and disseminate all-source finished intelligence; INR (unlike the CIA and DIA) has no dedicated collection capabilities.[28] The Office of Energy Intelligence is the intelligence arm of the Department of Energy and reports on a variety of topics. The Office of Intelligence Support is the treasury's intelligence office, which reports on international financial matters.[29]

The Federal Bureau of Investigation

The FBI is a law enforcement agency responsible to the Department of Justice. Its sole function within the Intelligence Community is counterintelligence. The assistant director of the FBI's National Security Division is the senior official responsible for U.S. counterintelligence activities within the United States. Coordination of U.S. counterintelligence activities overseas is the responsibility of the DCI until a matter has become a formal law enforcement investigation, when lead responsibility shifts to the FBI.[30]

The Intelligence Budget

The intelligence budget is divided into three major parts. They are the National Foreign Intelligence Program (NFIP), the Joint Military Intelligence Program (JMIP), and Tactical Intelligence and Related Activities (TIARA).

The NFIP includes the funding for the CIA and the national foreign intelligence or counterintelligence programs of the State Department, the Defense Intelligence Agency, the National Security Agency, the National Imagery and Mapping Agency, the National Reconnaissance Office, the army, navy, and air force, the FBI, the Department of Energy, and the treasury.[31] The NFIP is the only part of the intelligence budget that is directly under the purview of the DCI. Most of the NFIP also falls within the defense budget.[32] In Congress, the House and Senate Intelligence Committees have jurisdiction over the NFIP budget, subject to review by the House National Security Committee and Senate Armed Services Committee, respectively.

The JMIP comprises defense intelligence elements that support Defense-wide, multiple-service, or theater-level needs.[33] The JMIP is a budget category introduced in 1994, mostly consisting of programs formerly in TIARA.[34] The JMIP is developed by the Defense Department and falls under the responsibility of the deputy secretary of defense.[35] The House Intelligence Committee has jurisdiction over the JMIP. In the Senate, jurisdiction over the JMIP is held by the Senate Armed Services Committee, but the Senate Intelligence Committee staff is allowed to participate in staff-level Armed Services Committee meetings, and the chairmen and ranking minority members of the two committees have the opportunity to consult.[36]

TIARA comprises military intelligence assets and activities specific to a particular service that support combatant commands in that service. The TIARA budget is developed by the military services, and as such is an aggregation of service-specific programs. The House Intelligence Committee has jurisdiction over TIARA, but in the Senate, jurisdiction is in the hands of the Senate Armed Services Committee with the Senate Intelligence Committee staff allowed to participate, analogous to the way that the JMIP is handled.

Notes

Introduction

1. H. O. Yardley, *The American Black Chamber* (New York: Balantine, 1981; Bobbs and Merrill, 1931) Yardley himself was a remarkable cryptanalyst and intelligence officer who organized and directed army signals intelligence from its inception in World War I until its temporary demise in 1929.

2. For details on the creation and organization of the OSS, see John Ranelagh, *The Agency: The Rise and Decline of the CIA* (New York: Simon and Schuster, 1986), 47–92.

I
Why Intelligence Reform?

1. Presidential Commission on the Roles and Capabilities of the United States Intelligence Community [hereinafter the Aspin-Brown Commission], *Preparing for the 21st Century: An Appraisal of U.S. Intelligence* (Washington: U.S. Government Printing Office, 1996).

2. See *Making Intelligence Smarter: The Future of U.S. Intelligence* (New York: Council on Foreign Relations, 1996); *The Future of U.S. Intelligence* (Washington: Consortium for the Study of Intelligence, 1996); *In From the Cold: The Report of the Twentieth Century Fund Task Force on the Future of U.S. Intelligence* (New York: Twentieth Century Fund Press, 1996).

3. *IC21: Intelligence Community in the 21st Century,* staff study, House Permanent Select Committee on Intelligence (Washington: U.S. Government Printing Office, 1996). See also *IC21: The Intelligence Community in the 21st Century, Hearings of the House Permanent Select Committee on Intelligence,* May 22–December 19, 1995 (Washington: U.S. Government Printing Office, 1996).

Congress, as part (Title VIII) of the Fiscal Year 1997 Intelligence Authorization Act (Public Law 104-293—Oct. 11, 1996), passed the Intelligence Renewal and Reform Act of 1996 (50 USC 401 note). Some of its provisions reflect the recommendations of IC21. The act created some new offices in the Office of the Director of Central Intelligence (DCI) and formalized some of the DCI's authorities. The act did not, however, change the basic structure of the Intelligence Community, so its changes, while probably beneficial, must be regarded as largely cosmetic.

2
Essential Dogma and Useful Buzzwords

1. See, for example, Dan Balz, Bob Woodward, and Jeff Himmelman, "Afghan Campaign's Blueprint Emerges," *Washington Post,* 29 January 2002, which makes the CIA look like the lead agency in the war.

2. *IC21* (p. 76) calls for giving the DCI limited authority to reprogram funds within the National Foreign Intelligence Program. The Intelligence Renewal and Reform Act of 1996 (Public Law 104-293–11 October 1996) requires the reporting to the DCI and the secretary of defense of budget execution data for all intelligence activities (sec. 807).

3
Making Dollars Yield Useful Intelligence

1. This informal "concurrent" authority of the DCI over high-level appointments was formalized in law in 1996 in the case of the appointments of the directors of the NSA, the National Imagery and Mapping Agency, and the National Reconnaissance Office. The DCI is also to be consulted regarding appointments of the director of the DIA, the assistant secretary of state for intelligence and research, and the director of the Office of Nonproliferation and National Security of the Department of Energy. Public Law 104-293–11 October 1996, section 815.

4
The World of Military Intelligence

1. One Department of Defense organization, the Defense Airborne Reconnaissance Office, is not treated in this book. It is a relatively new organization intended to overcome the fragmentation in the development of aerial reconnaissance platforms among the military services and the Intelligence Community. In principle it is likely to create some of the same problems generated by the National Reconnaissance Office, but that remains to be seen. Because no obviously more effective alternative structure is easy to conceive for this troublesome area of research and development and resource management, I have omitted it from this analysis.

2. See Dan Elkins, *An Intelligence Resource Manager's Guide* (Washington: Defense Intelligence Agency, Joint Military Intelligence Training Center, 1994), 100–101. Thanks to Mr. Elkins for reading one of the drafts of this book and making many helpful comments.

3. Elkins, pp. 38, 41, 42.

5
Listening to Learn: Signals Intelligence

1. The NSA also has responsibility for a large nonintelligence function. It manages and develops the national cryptologic system—that is, all cipher devices and codes. It performs this service not only for the Defense Department but also for all federal agencies which need secure communications. This function lies outside the bounds of this study.

2. The NRO continues to mislead intelligence users and the Congress by marketing itself as an intelligence producer. See prepared testimony by NRO Director Keith Hall before the Senate Armed Services Committee, 12 March 1997, p. 4.

3. 18 U.S. Code 798, 1951.

4. James Bamford, *The Puzzle Palace: A Report on America's Most Secret Agency* (New York: Houghton Mifflin, 1982). The director of the NSA at the time made a considerable effort to get the attorney general to take legal action against the author because the disclosures were so egregious. When Bamford published a second book, *Body of Secrets: Anatomy of the Ultra-Secret National Security Agency* (New York: Doubleday, 2001), he was given a book party by the NSA, an index of the recent change in the agency's media policy. Although Bamford has consistently denied that he had access to inside sources, the NSA has his signature on a nondisclosure statement from the time he worked there as a navy enlisted man from the Naval Security Group. In the statement he acknowledged that he understood that he was subject to 18 U.S. Code 798 in the event that he revealed the content of signals intelligence or information about how it is collected.

6
Looking to See: Imagery Intelligence

1. See U.S. Congress, Senate, Committee on Armed Services, Hearings, *Operation Desert Shield/Desert Storm* (Washington: GPO, 1991; S. Hrg. 102-326), pp. 320–321.

7
Spying to Know: Human Intelligence

1. See John M. Newman, *JFK and Vietnam: Deception, Intrigue, and the Struggle for Power* (New York: Time Warner, 1992), for a well-documented account of this example.

8
Spying on Spies: Counterintelligence

1. Harvey Klehr, John Earl Haynes, and Fridrikh Igorevich Firsov, *The Secret World of American Communism* (New Haven: Yale University Press, 1995); Harvey Klehr, John Earl Haynes, and Kyrill M. Anderson, *The Soviet World of American Communism* (New Haven: Yale University Press, 1998).

2. John Earl Haynes and Harvey Klehr, *Venona: Decoding Soviet Espionage in America* (New Haven: Yale University Press, 2000).

3. See Jerrold Schecter and Leona Schecter, *Sacred Secrets* (Washington, D.C.: Brassey's, 2002). This book contains an NKVD document naming Oppenheimer as helpful to the Soviet effort to gain information about the Manhattan Project.

4. For some "how to" cases, including Clyde Lee Conrad, see Stuart A. Herrington, *Traitors Among Us: Untold Stories of the Cold War* (Novato, Calif.: Presidio, 1999). Herrington, a retired army colonel who finished an illustrious career as the army's top counterintelligence operational officer, began his outstanding work in Vietnam, uncovering Vietcong and North Vietnamese agents. For that period, see his *Silence Was a Weapon: The Vietnam War in the Villages* (Novato, Calif.: Presidio, 1982). For evidence of weaknesses in army counterintelligence that allowed Conrad and others serving in Germany to conduct espionage in the first place, see John O. Koehler *Stasi: The Untold Story of the East German Secret Police* (Boulder, Colo.: Westview, 2000), which is based on abundant sources from former East German intelligence officers who penetrated the West German government as well as the U.S. military there.

5. Herrington, *Traitors Among Us*.

6. See, for example, Robert Lamphere and Tom Shachtman, *The FBI-KGB War: A Special Agent's Story* (New York: Random House, 1986), and David C. Martin, *Wilderness of Mirrors* (New York: Ballantine, 1980).

Appendix

1. See CIA, Public Affairs Staff, *A Consumer's Guide to Intelligence* (Washington, D.C.: CIA, document number PAS 95-00010, July 1995), p. vii.

2. CIA, *Consumer's Guide,* p. 1.

3. Aspin-Brown Commission, p. 48.

4. Public Law 104-293–11 October 1996, 110 stat. 3477–3478.

5. CIA, *Consumer's Guide,* p. 41; see also *IC21,* p. 74.

6. Aspin-Brown Commission, p. 56n.

7. Mark M. Lowenthal, *U.S. Intelligence: Evolution and Anatomy* (Westport, Conn.: Praeger, 1992), p. 112.

8. CIA, *Consumer's Guide,* p. 15.

9. CIA, *Consumer's Guide,* p. 56; Lowenthal, *U.S. Intelligence,* p. 133.

10. CIA, *Consumer's Guide,* p. 15.

11. Lowenthal, *U.S. Intelligence,* pp. 132–33.

12. CIA, *Consumer's Guide,* pp. 41, 56.

13. CIA, *Consumer's Guide,* p. 56.

14. Lowenthal, *U.S. Intelligence,* p. 111.

15. Lowenthal, *U.S. Intelligence,* p. 111.

16. Lowenthal, *U.S. Intelligence,* pp. 18–19; in 1992 Congress amended the National Security Act specifically to authorize the CIA to collect human intelligence and to provide overall direction for human intelligence collection by other U.S. government agencies. Aspin-Brown Commission, p. 61n.

17. Aspin-Brown Commission, p. 62.

18. CIA, *Consumer's Guide,* p. 23.

19. Aspin-Brown Commission, p. 132.

20. National Imagery and Mapping Agency, booklet, "NIMA Establishment Ceremony" (Fairfax, Va.: NIMA, 1996), pp. 3, 11.

21. National Imagery and Mapping Agency, Office of Congressional and Public Liaison, media release (fax), 1 October 1996.

22. NIMA's first and current head, Rear Adm. J. J. Dantone, a two-star admiral, was appointed as acting director.

23. Brig. Gen. Robert (Rick) Larned, USAF, director of imagery systems acquisition and operations, NRO, briefing slides presented to the Society of Old Crows, Alexandria, Va., 2 October 1996 (fax).

24. CIA, *Consumer's Guide,* p. 10.

25. CIA, *Consumer's Guide,* pp. 11–12.

26. Aspin-Brown Commission, p. 109.

27. The designation J-2 for the joint staff (G-2 or S-2 in army staffs) signifies the intelligence function and staff size. Staff functions are numbered from 1 to 5. Personnel is 1; intelligence, 2; operations, 3; supply, 4; and civil affairs, 5. Each function is served by a staff, the size of which varies depending on the size of the unit.

28. CIA, *Consumer's Guide,* pp. 24–27.

29. CIA, *Consumer's Guide*, p. 27.

30. Aspin-Brown Commission, p. 58.

31. *IC21*, p. 11.

32. *IC21*, pp. 12, 71.

33. *IC21*, pp. 11, 72.

34. Department of Defense Directive, 7 April 1995, no. 5205.9, Joint Military Intelligence Program (JMIP) (fax). See also Richard A. Best, Jr., *Intelligence Issues and the 104th Congress* (Washington: Congressional Research Service, 26 September 1996, Order Code IB95018), p. CRS-6.

35. See U.S. Congress, Senate, Select Committee on Intelligence (SSCI), Authorizing Appropriations for Fiscal Year 1997 for the Intelligence Activities of the United States Government and the Central Intelligence Agency and the Central Intelligence Agency Retirement and Disability System and for Other Purposes (Washington: Government Printing Office, Report 104-258), p. 2.

36. SSCI, p. 3.

Index